To Kay

Good luck with
your tomato patch.

Walter D Culams
10/06/2011

The Texas
Tomato Lover's
Handbook

Texas A&M System

**AGRILIFE RESEARCH
AND EXTENSION SERVICE
SERIES**

Craig Nessler and
Edward G. Smith,
General Editors

The Texas Tomato Lover's Handbook

William D. Adams

PHOTOGRAPHS BY **WILLIAM D. ADAMS**

AND **DEBORAH J. ADAMS**

TEXAS A&M UNIVERSITY PRESS COLLEGE STATION

Library of Congress Cataloging-in-Publication Data

Adams, William D., 1946–

   The Texas tomato lover's handbook / William D. Adams ;
photographs by William D. Adams and Deborah J. Adams.—1st ed.

      p. cm.—(AgriLife Research and Extension Service series)

   Includes index.

   ISBN 978–1–60344–239–8 (pb-flexibound : alk. paper)

1. Tomatoes—Texas—Handbooks, manuals, etc.

2. Tomatoes—Breeding—Handbooks, manuals, etc.

3. Tomatoes—Diseases and pests—Control—Handbooks, manuals,
etc. 4. Tomatoes—Texas—Pictorial works.

I. Title.   II. Series: AgriLife Research and Extension Service series.

 SB349.A253   2011

 635.'64209764—dc22

 2010029555

General editors for this series are Craig Nesslor, director of Texas
AgriLife Research, and Edward G. Smith, director of the Texas
AgriLIfe Extension Service.

Contents

Preface

In the mid-1990s, I launched a Texas Cooperative Extension educational program (now Texas AgriLife Extension Service) called Team Tomato in Harris County (Houston, Texas) with a lot of help from extension staff, including Tom LeRoy, Robert "Skip" Richter, Sam Cotner, Bart Drees, and Jerral Johnson. Volunteers, including George and Mary Stewart, Wendell and Barbara Biggers, and Tom Robb, also carried the banner. That first year we signed up close to five hundred participants in three locations for a day-long program on nothing but tomatoes. Not one sentence was devoted to bush beans, collard greens, or sweet corn—we did cover some close relatives like peppers and eggplants, but the star of the show was the tomato.

It was rather obvious that tomatoes were a big draw, and frankly we love tomatoes, so it wasn't hard to take the passion and run with it. The Extension test garden in Harris County was home to twenty or more new varieties of tomato each year for the last decade of my career in Extension Service. Since my retirement to the Burton area, our oversized kitchen garden has seen a similar emphasis on tomatoes. At this point it would be tempting to settle on a couple of favorites like Champion, Viva Italia, and Sungold; reduce our plant numbers; and stop this quest for better-tasting, pest-resistant tomatoes, but that's no fun. There are currently more than forty new tomato varieties plus a few old favorites seeded in our small greenhouse for the coming season. Apparently we are already working on a supplement to this text. Right now I wish I had a couple more years of testing before finalizing this book. The reality is you can never catch up because so many new varieties come out each year. We will leave some blank pages for notes so you can keep track of your successes.

Master Gardeners have been such an important part of my career, I could not have begun to learn what I have about tomatoes without their help and support. They made our test gardens in Harris County a mini experiment station with their hard work and dedication. And Master Gardeners throughout the state of Texas are doing likewise. If you have a tomato question, ask a Master Gardener. In particular, I want to thank Master Gardener Clyde Cannon for his Photoshop tutoring and Cindy Appleman for her careful and thoughtful review of the manuscript.

I owe Shannon Davies special thanks for her concept of a book dedicated to home gardeners who love the wonderful freshness of a vine-ripe tomato, and especially for her confidence in me to put it on paper.

The "we" in this introduction includes my wife, Debbi. She's not an avid gardener although she loves plants, but she is a patient and understanding partner who reads my ramblings from a beginning gardener's perspective. Things I presume readers know, she points out need more explanation. If there is a photo of me in the book, she took it. Photography has long been a passion of mine, but Debbi is also a pro. If the tomatoes needed to be covered in the middle of the night because a late frost threatened, she was there. For these and so many other reasons, this book is dedicated to Debbi.

The Texas
Tomato Lover's
Handbook

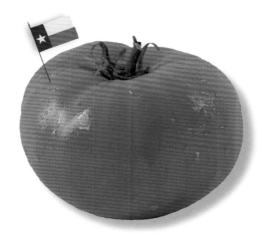

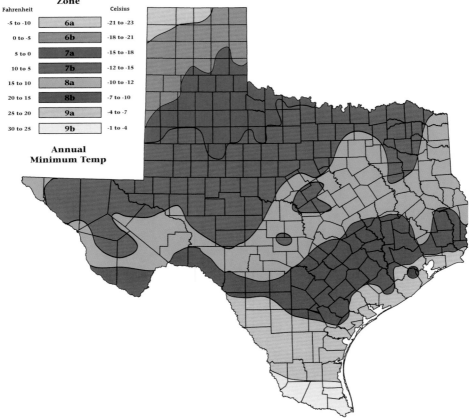

USDA Hardiness Zone

Fahrenheit		Celsius
-5 to -10	**6a**	-21 to -23
0 to -5	**6b**	-18 to -21
5 to 0	**7a**	-15 to -18
10 to 5	**7b**	-12 to -15
15 to 10	**8a**	-10 to -12
20 to 15	**8b**	-7 to -10
25 to 20	**9a**	-4 to -7
30 to 25	**9b**	-1 to -4

Annual Minimum Temp

History of the Garden Tomato

Did the tomato come over with an Italian immigrant planning to open a pizza parlor featuring a wonderful tomato sauce? No—the tomato is actually a New World species from the mid- to high-altitude regions of Peru and nearby regions of South America. The tomato wasn't delivered to the European continent until the late fifteenth or sixteenth century (Columbus or Cortés—depending on whom you believe). To their credit, the Spanish and Italians took to the tomato as a food source almost immediately, and by the mid-sixteenth century it was widely cultivated in the Mediterranean region. The British considered tomatoes (which they also called "wolf peach" or "love apple") poisonous until the mid-eighteenth century, and in view of the number of toxic plants in the family Solanaceae, perhaps their caution was justified—one wouldn't want to consume tomato foliage today.

The tomato was widely spread in South America and Mexico, likely as a small, yellow fruit long before our European ancestors got their hands on it. The Aztecs were even making salsa before the Spanish arrived in Latin America.

The American colonies remained cautious, and true to their—at the time—British heritage, were slow to set the stage for the first pizza parlor. Thomas Jefferson had eaten tomatoes in Paris and began cultivating them in his Monticello garden in the late eighteenth century. Most colonial gardeners, however, remained skeptical until the early nineteenth century. In 1820, Colonel Robert Gibbon ate a basketful of raw tomatoes on the steps of the Salem, New Jersey, courthouse, and much to the horror and amazement of the onlookers, he survived, and the popularity of the tomato was assured. By the late 1800s Henry Heinz was selling ketchup.

FRUIT OR VEGETABLE?

Actually, it's both. Botanically the tomato is a fruit, but from a culinary standpoint it is used as a vegetable.

Tomato varities from the small Sungold to the standard-sized Bush Champion.

2 Must-Have Tools for the Tomato Patch

The right tools can make tomato production much easier. Instead of using a tiller, consider a good stainless-steel spading fork and a small flat-edged shovel. For moving compost around, you need a hay fork. Trowels are an absolute necessity—the long-handled kind can save you a bit of back pain. Hand cultivators and the like come in handy, and don't forget the knee pads. Who says gardening isn't a contact sport?

One of the most important tools you need in the tomato patch is a good sprayer. Even if you grow organically, you will need to spray organic pesticides, compost tea, and organic fertilizers. Eventually you will want to own at least two sprayers—one for weed-control and one for pesticides and fertilizer. A hose-on sprayer can fill the bill in some cases, but spraying is hard to do without a good pump-up pressure sprayer. A backpack sprayer for the big jobs and a quality 1½- to 2-gallon hand- or shoulder-carried sprayer should cover most of your spraying chores.

Every gardener needs a good hoe. One of my favorites is the scuffle hoe. Its triangular shape works with a forward push—actually, it probably works best if you are backing up through the weed patch. Since it

Master Gardener Kelly Reed sprays tomatoes at the Montgomery County test garden.

Hoes come in a variety of styles and shapes. You will probably want more than one style. A scuffle hoe for quick weed removal and a standard hoe for seed trenches will usually get the job done.

A scuffle hoe is handy for weed cleanup. It is used with a pushing action and cuts the weed roots off shallow enough that most tomato roots are left undamaged.

makes a shallow cut into the soil, you reduce the risk of damaging tomato roots, and you can get rather close to the tomatoes if you are careful. Stirrup hoes look just like a stirrup—the bottom edges are sharp, and the hoe can be used with a push or pull motion. It also makes a shallow cut into the soil and causes less damage to crop roots. A chopping hoe is handy for making planting trenches and for attacking overgrown weeds. Like most tools, hoes need to be kept sharp with a file. Don't forget a rake—you'll need it to pull up the soil into ridges.

You can expect to use a lot of plastic tie tape securing the plants to the cages and redirecting their errant ways. Fiber row cover can be used to get the tomato plants off to a good start by wrapping it around the cages to protect the plants from cold and drying winds. It can also be used as a wrap around tomato clusters to keep the bugs and birds away if you're trying to grow the perfect red cluster to show your friends and neighbors—and especially if you're a photographer. You will also need a pocketful of clothespins.

Quality pruning shears will come in handy, especially when tackling the old tomato vines that need to be cut out of the cages before they can be removed from the garden and stored. Choose the scissors type, and keep them clean and sharp. Keep a sharp kitchen or melon knife handy for taste testing.

Make your own labels (I've used old window shade slats), and save your grocery bags—you will need them when you harvest the tomatoes. Distinguish varieties by picking into separately labeled harvest bags or writing on the tomato shoulders with a marking pen. If you're going to brag about your tomato crop, at least call them by name! It's hard to beat a #2 pencil for writing on labels—graphite does not wash off or fade in the sun. You will still need to invest in a marking pen to write on the harvest bags.

Don't forget gloves. You need a good work pair for pruning and picking to limit the green staining on your hands. Heavy-duty rubber gloves are advisable even if you are only spraying organic pesticides. Wear them when you work with fertilizers, too. Chemical fertilizers will have a drying effect on your hands, and organic fertilizers usually smell strong and may harbor disease organisms.

Be creative! Gardeners like to invent their own tools and garden devices. The *cantarito* is an Incan pot used to water a 3-foot-diameter section of the garden—one between each tomato plant should suffice. This unglazed ceramic "genie jug" is partially buried and filled with water. Every few days, or every day in hot weather, you fill it again, and the water gradually seeps out into the root zone. These might be a bit expensive for a big patch, but for years gardeners have been burying gallon cans with holes punched in the sides (at the bottom) for water reservoirs between their tomato plants.

To keep track of your tomato varieties, pick them into individually labeled bags or use a marker to write the name on each tomato. Then you can taste-test them and be sure you grow the ones you like.

Home gardeners can use various methods in their attempt to prevent bird damage on their tomatoes. You can paint old baseballs red or get red Christmas ornaments and hang them in the plants as the tomatoes begin to ripen. Birds will peck them and leave your tomatoes alone . . . well, it makes a good story. Other gardeners place shallow containers filled with water around the tomato patch, in the theory that the birds just want water. Personally, I think they would eat my BLT if I left it unattended in the garden.

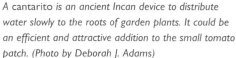

A cantarito *is an ancient Incan device to distribute water slowly to the roots of garden plants. It could be an efficient and attractive addition to the small tomato patch. (Photo by Deborah J. Adams)*

(top right) Many a Texas tomato has been planted using this system—cans for water and fertilizer, black plastic for weed control, and a can (minus its bottom) around the plant.

(middle right) Bird damage shows up in tomatoes as deep holes pecked into the fruit.

(bottom right) After trying to peck a baseball painted red, most birds give up.

A Recipe for the Perfect Tomato Crop

Basic Cultural Needs

3

The homegrown tomato must have outstanding flavor. The texture should be firm, not hard, and never grainy or mushy. Tomatoes should be juicy with zingy tartness and complex, sweet tomato flavors. Anything less is unacceptable and should never be considered regardless of other traits the variety may possess, including disease resistance, health and vigor, beautiful color, uniform shape, shipping qualities, heirloom mystique, or the numerous other "beauty queen" factors that may be touted. The ideal tomato is typically medium to medium-large in size. Although there is certainly room for the small to cherry-sized tomato, oversized "beefsteak"-type tomatoes are rarely productive and often unattractive. It's not uncommon to end up with one slice in the middle after cutting all the folds and green core out of these monsters.

Adapted varieties should yield 40 to 50 pounds of mostly edible tomato fruits per plant. Gardeners can certainly disagree about their favorite varieties, and the plethora of new varieties offered each year can be staggering, so don't hesitate to add notes on your successes and failures each season. Labeling plants and keeping good notes about the garden are the only ways to determine the best varieties for your garden and table. One more important fact—beware of the garden expert who doesn't have a garden, especially one who doesn't enjoy eating fresh tomatoes. In the Adams household, we don't just grow tomatoes; we lust for them fresh from the garden. Keep the following parameters in mind as the adventure begins.

LOCATING THE TOMATO PATCH

Tomatoes need full sun or a minimum of six to eight hours of full sun to produce good crops. They appreciate some shade in midsummer, especially in late afternoon, but this is a low-productivity time anyway. Most gardeners give up on tomatoes in mid- to late summer and begin thinking about the fall garden while fishing in Alaska or strolling through a cool mall. Keep enough hot-weather varieties or cherry types for use with the okra or in salads, but for most Texas gardeners late summer and fall are not bonanza times for tomatoes.

Tomatoes need good drainage. For some gardeners, this means raised beds or containers. For others, the way to go is to work the soil up into ridges and plant on the ridge.

It's also important to locate the tomato patch away from trees and shrubs that will compete for water and nutrients. In some cases these competitors cannot be avoided. However, by building raised beds, you can at least keep the roots out of the tomato bed for a while. If possible, locate the tomatoes close to the house just for convenience.

GROWING YOUR OWN TRANSPLANTS

Most home gardeners buy plants at a local nursery. Fortunately, there are more varieties available as transplants than ever before—from hybrids to heirlooms—but if you want the latest hybrid or the lesser-known heirloom, you will need to grow your own. Building a home greenhouse is the best answer to growing your own plants, but you can make do with a cold frame or hot bed. Growing a few plants indoors under fluorescent

The first extension to the original Adams kitchen garden allowed us to grow more tomatoes.

lights is another solution. Just be sure the lights are close to the leaves—6 inches or closer—even if you have to prop flats up on overturned pots or mini-benches made of 1 x 4s and a couple of bricks.

Tomato seeds for transplants are usually started six to eight weeks before time to transplant, which for most gardeners is two to four weeks after the average frost date in their area.

If you're willing to take a chance, then you might set out transplants two weeks before the average last frost date. Just be prepared to cover

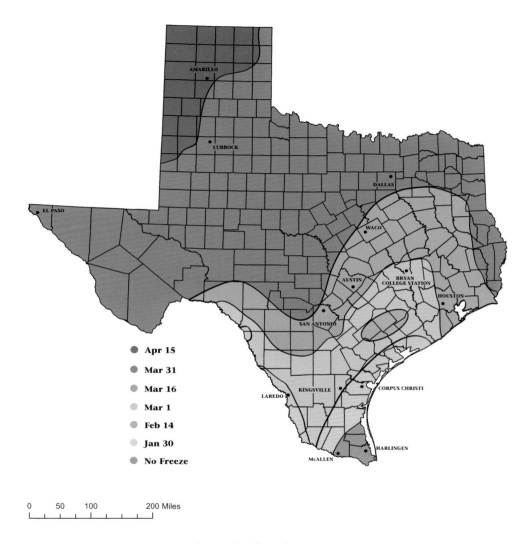

Average Last Spring Frost Dates

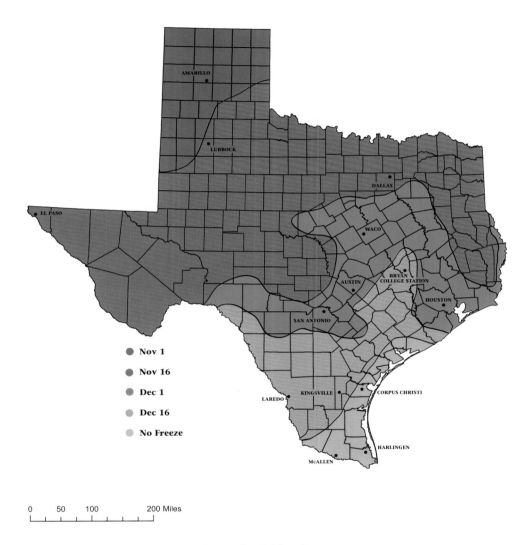

Nov 1
Nov 16
Dec 1
Dec 16
No Freeze

0   50   100   200 Miles

Average First Fall Frost Dates

Cold frames have been used for hundreds of years to get plants through cold weather.

Tomato seedlings need to be close to the fluorescent lights—six to twelve inches will ensure that they develop dark green and stocky.

them if a freeze is predicted. However, even seedlings started on a sunny, protected patio after the danger of frost often produce well and outlast the "jump-the-gun" gardeners with their early spring harvest. The heat, humidity, and pests of summer really do challenge tomatoes, but plants set later seem to hang on longer with a bit of extra care. Fall tomatoes are often a bust, but if you must try, plan on a May/June seeding date, setting plants in mid-June to late July (five to six months before the average first fall frost date). This will likely require some special attention to get these plants established during the heat of summer. Cover the row with fiber row cover or nursery shade cloth, and water often to prevent stress.

Resist the temptation to make your own potting soil from backyard dirt and peat moss to start your seedlings. Buy a soilless mix from a local nursery to enhance your chances of success. You can purchase seed flats or make some from frozen-dinner trays. Foam coffee cups with holes in the sides at the base make good growing containers and can be used for starting seeds. Place them in a location with warm bottom heat, like the top of the refrigerator, or purchase a heating mat made especially for starting seeds. Don't even think about using the heating pad tucked into a bathroom closet, because you need a shockproof device designed to be used around water.

Plant the seeds in a shallow trench made in the potting soil with a wooden spatula or something similar—about ¼ inch deep is sufficient—and cover them with more potting soil (about four times the seed's thickness) or use vermiculite. The sterile vermiculite can help reduce seedling diseases. Finally, set the flats in a tray with warm water to soak the seeds from the bottom and ensure that the soil mix is evenly moist. Set the seedling tray aside to drain thoroughly, and then slip it into a plastic bag to guarantee high humidity. You need to watch carefully and remove the tray from the bag as soon as the seedlings germinate; otherwise, they may melt away from seedling diseases like damping-off.

To move the seedlings from the seed flat to individual containers, first dampen some potting soil and fill peat pots or foam coffee cups with soil

COLD FRAME / HOT BED

A cold frame or hot bed can be a lot cheaper than a greenhouse, and both are capable of producing a lot of transplants. A cold frame is simply a box, generally with a wood frame and slanted, hinged lid covered with glass, fiberglass, or plastic. For the most part a hot bed is just a cold frame with some provision for heat, such as a heating cable designed for outdoor use placed under the root zone to provide bottom heat or a subsurface layer of fresh manure to produce heat from oxidation. The hot bed is typically used to start seedlings because you can control the soil temperature for good germination (a soil temperature of 70 to 75 degrees Fahrenheit is required for most warm-season plants like tomatoes). Then it can function as a cold frame a few weeks before transplanting to keep the plants cooler and dryer before moving them to the cold, cruel world of the garden. The cold frame relies on the "greenhouse effect" from the glass or fiberglass cover to build up heat during the day, which then gets the plants through the colder nighttime temperatures. On really cold nights you can cover the cold frame or hot bed with an old blanket or rug for better insulation. Either is best located where it will face south for maximum light and heat absorption. The daytime temperature can be just as critical as the nighttime temperature, so it will be important to adjust the cover to allow excess heat to escape on warm, sunny days. The cover is hinged at the top so it can be lifted and propped open at varying heights to adjust the temperature inside.

Heating cables and cold frame/hot bed kits are available from a number of seed catalogs and greenhouse supply companies. Just do an Internet search for "cold frame" or "hot bed," and you'll find both in abundance. See Ohio State Extension bulletin HYG-1013–88 for a sample plan. University of Missouri bulletin G6965 is even better. If you're handy with woodworking, building your own could be a fun project. Plan to slant the roof at a 15- to

30-degree angle (you need a slope of 1 inch per foot of back to front width), and consider building the sides with 2 x 8s, 2 x 12s, and so on, since they will be stronger and provide more insulation than boards 1 inch thick. A 3- x 5-foot hot bed will start a lot of transplants.

mix (remember to poke holes in the sides at the base of the cup by using, for example, a sharp pencil). Put a label in each container with the variety name, or write on the side of the foam cups. Then use a pencil or any sharpened stick, like a piece of bamboo, to make a hole in the soil to receive the seedling roots. The technical term for this useful tool is a dibble.

When the seedlings' first true leaves begin to appear, use a pencil or old knife to lift the plant under the roots, and by holding on to the cotyledon (one of the first two seedling leaves), carefully lift the seedling out and separate it from its neighbors. Don't grab the stem. If you damage the stem, the seedling will die or be stunted. If you damage a bit of the cotyledon, the plant will recover virtually unhindered. Seedlings often have more roots than will easily fit into a pencil-sized hole, so it's okay to pinch off some of the lower portion of the roots. Resist the temptation to wind and shove the whole root mass into the opening because seedlings recover faster with pruned roots that are not cramped in the hole. Water the seedlings in immediately—at least after each flat—even though the potting soil is moist. You must settle the soil around the roots for them to efficiently take up moisture.

A few days after the transplanted seedlings have settled in, begin using a dilute fertilizer solution—about ½ to ⅓ strength soluble fertilizer like 20–20–20 or an organic fertilizer like liquid seaweed or fish emulsion. Applying a liquid fertilizer once a week should keep the seedling plants

Seedling flat with commercial potting soil.

Place a seedling flat in a plastic bag to ensure constant moisture as seeds begin the germination process. Be sure to remove the flat from the bag as soon as seedlings emerge.

growing actively, but you don't want to overdo it. Plants can quickly stretch too tall, and you will have to lay them down in a planting trench when you set them out. However, in most cases, any stretching is caused by low light conditions, so get them close to fluorescent lights or in full sun with a cold frame or greenhouse.

An alternative technique involves planting two to three seeds in each cell of a seedling flat or in individual containers and then thinning the small seedlings to one per cell before they develop the first true leaves. This may waste some plants, but it is a quick way to start a few plants of a number of varieties.

Seedling flats in a sunny south window.

Use a dibble, such as a sharpened pencil or dowel, to make holes for the tomato seedlings.

When transplanting tomato seedlings, handle the plant by a leaf, not the stem.

Direct seeding to individual pots or cells in a planter flat is an alternative to seeding and transplanting. Be sure to thin to a single plant while the seedlings are still small.

GRAFTING TOMATOES

You have likely seen ads for the grafted tomato/potato with a tomato top and potato roots. The theory is you get potatoes in the soil and tomatoes aboveground. Unfortunately, you get very few of either. Potato roots do not support a tomato top very well, and the potatoes need to be harvested long before the tomatoes are ripe. To get the potatoes without destroying the tomato plant necessitates a subterranean "snatch and grab," or you'll have to dig the potatoes and forget about the tomato crop (you were going to harvest only a few anyway). It makes a great ad for the Sunday paper, but it's a bad idea in the garden.

So why are horticulturists still experimenting with tomato grafting? A vigorous, healthy rootstock with resistance to nematodes and other root problems can give a "leg up" to heirloom or otherwise puny, disease-challenged varieties and make them more productive. In theory, Texas gardeners might be able to grow Brandywine tomatoes, which prefer a New England summer, by using a rootstock with better pest resistance and increased vigor. Greenhouse tomato growers have been using grafted tomatoes for a while, and field production looks more promising. There are even commercial varieties selected for rootstock, or for fun you could just pick out a hybrid variety with vigor and pest resistance and use that for a rootstock in your heirloom grafting experiment. It should be noted that any foliar disease resistance that the rootstock might have does not translocate to the tops of your heirloom. The only benefit is from a healthy rootstock.

Tomato plants are always succulent, so there is no dormant phase as in fruit trees. This means that keeping the grafted plant alive while it heals and adjusts to the great outdoors is a bit tricky. Plants used for grafting need to be about the same size—usually 4 to 6 inches tall—and the stems should be the same diameter where they are cut and combined. Some sources recommend a 45-degree cut, but a 90-degree cut is easier to match. Special clips are used to keep the cut stems together (available from source no. 3 on the Tomato Source List). First, cut the rootstock plant and place a clip on the stem with half of the clip open (unfilled) to hold the top. Plastic tubing with an inside diameter the same size as the tomato stems and slit lengthwise on one side so that it will expand and fall off after the graft heals can also be used. Next, cut the top (this will be your heirloom or weakling) at about the same diameter and slip it into the clip. (This top piece is called a scion when referring to grafts with fruit trees or other woody species.) Be sure to push them together—they won't jump the gap. And don't worry about removing the clip, as it will fall off as the tomato stem grows. Slip the entire grafted plant into a zip-top plastic bag and seal it. Be sure the bag is large enough that you don't have to bend the plant to get it into the bag. Now find a dark corner in the bathroom or an out-of-the-way location, and leave the bag for a few days to start the healing process. The rootstock should be watered before starting; in a few days you may need to add some water to the bag.

A razor blade is used to cut the stems of tomato plants for grafting. Though a 45-degree cut gives more area for contact, a straight cut at 90 degrees is easier to make and the sections fit nicely in the clip or tubing.

The top portion of the graft (scion) is inserted into a small plastic tube or clip that was first placed on the rootstock. This holds the tissues together as they heal.

Total darkness isn't necessary, and a little room light should help keep the chlorophyll green—though keep the light low to reduce demand on a temporarily weakened system. Gradually move the plant to a brighter location, open the bag for a few hours each day, and eventually leave the bag open in a bright window.

If you have a greenhouse or cold frame, you can begin to transition the plant to more sunlight and the open atmosphere. If you're working with a number of plants in the greenhouse, it will help to construct a small shade cover over the plants and keep them misted to ensure high humidity. This greenhouse humidity chamber will negate the use of a plastic bag. Acclimating a few plants to the outdoors from a warm, humid bath-

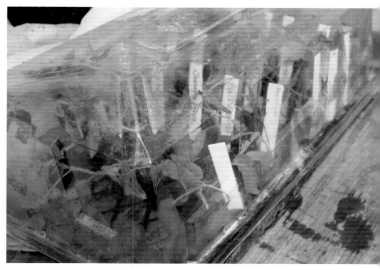

Grafted tomatoes ready to be placed in a plastic bag to keep them from wilting severely and speed the healing process.

A tomato graft transition chamber.

A tomato graft transition chamber with the lid lifted.

Healed tomato grafts with and without clip.

A grafted tomato plant will gradually push the clip off the tomato stem as it expands.

room in the spring can be a challenge. You may have to put them out for only a few hours each day when temperatures are the warmest.

When the cut tissues have calloused over and the stems appear strong, it is time to set the plants out. Grafting takes awhile, so plan on an additional two to four weeks to produce these grafted transplants compared to the time required for direct-seeded plants. Don't set the plant where the stem of the top portion touches the soil—if the top develops roots, you have lost all or most of the advantage of growing it on a more vigorous rootstock. A small bamboo stake can help keep the plant upright and off the ground where it might root. This is contrary to the popular technique of laying the stretched-out stem of an overgrown (but ungrafted) tomato plant in a trench so it will develop roots and take off faster. Remember that we are confining this plant to a better root system. You will still need to circle the stem with cardboard or aluminum foil at the soil line to discourage cutworms. To really evaluate whether grafting is worth the time and effort, plant a row of grafted heirlooms next to an identical row of heirlooms on their own roots and weigh the production of each.

PREPARING THE SOIL

The gardener with the most compost wins! Seriously, you need to be fanatical about organic matter. Make it, buy it, pick up the neighborhood lawn clippings—whatever it takes; this is a never-ending challenge.

To Till or Not to Till

At some point in any discussion of soil preparation, the subject of "to till or not to till" comes up. Most gardeners don't need a tiller. Even if you want one for the initial soil workup, you can rent one. You may find a mini-tiller useful—even with large raised beds. It can be used for shallow incorporation of compost and fertilizer and for taking out weeds, especially if you get after them while they are small. It's important to remember that you still can't get too close to the tomato plants without damaging a lot of roots. For most of us, a good selection of scuffle or stirrup hoes will do the job with more surgical precision and a little more effort.

Mushroom compost.

Importance of Organic Matter

A soil rich in organic matter and nutrients is essential for good tomato production. The interest in organic gardening has highlighted the importance of compost for amending soils, whether clay or sandy in texture. It's also a valuable component of soil mixes. Making your own compost usually results in the very best product, but most of us will have to buy at least some of it.

Most gardeners don't need a big tiller, but mini-tillers are handy for shallow incorporation of compost and fertilizers. They also make quick work of small weeds prior to planting or between rows early in the season.

My garden is almost pure mushroom compost. There is likely better compost, and in the future we may rely more on neighbors' manure piles, but for now mushroom compost is available by the dump truck load, and our tomatoes have thrived in it. Mushroom compost is variable depending on how long it has been sitting at the dirt yard, but tomatoes are very salt tolerant (salt content may be high in fresh mushroom compost) and should do fine in all but the freshest, steaming compost. As a minimum, incorporate 6 to 8 inches of compost worked into the tomato row or at least worked into the immediate root zone (24 inches around the planting hole). This should result in a thriving and beneficial microorganism population that will get plants off to a good start.

How to Recognize Good Compost

Since there is no legal definition of compost, you really have to check it out before ordering compost from the local dirt yard. Real compost should not be strong smelling but have a rich, earthy smell. Mushroom compost can be rather fresh and a bit odiferous in most cases, but it sweetens up with a bit of exposure and rainfall. Good compost should be moist but not wet, and most of the ingredients that went into its production will be broken down enough that they will be unrecognizable. Finding a bunch of wood slivers or stems—even if they are black—is a clue that the process is not finished. Additives like slag are sometimes incorporated by unscrupulous compost vendors to give organic matter a finished, black look. The slag adds alkalinity and toxins that don't benefit the soil or the gardener.

Can you make your own compost? Of course. That's the best way, but it does require some effort. Resist the urge to buy a plastic or metal compost bin. Folks typically fail with these devices, and building a homemade bin or two is not that hard. It helps to turn the compost every four to six weeks, and there is a benefit of having two—you start with one empty bin and turn the compost from the other one into it. A ratio of green material (grass clippings, manure, etc.) to brown material (leaves, wood chips, etc.) of 25 to 30 carbon to 1 nitrogen is about right. In the following ratios, carbon (brown material) is listed first followed by nitrogen (green material). Some of the materials that are high in carbon include news-paper 175:1, leaves 60:1, sawdust 325:1, and corn stalks 75:1. Materials comparatively high in nitrogen include alfalfa 12:1, cottonseed meal 7:1, food waste 20:1, manure 15:1, and vegetable scraps 25:1. If you don't have enough high-nitrogen organic material, you will need to supply nitrogen from an inorganic source like a 15-5-10 fertilizer or an organic source like blood meal or fish meal.

Compost piles are often described as a layer cake—organic matter, then soil, followed by more organic matter (usually with a cup or two of organic fertilizer such as blood meal or a commercial fertilizer like 15-5-10) added in. If most of the organic matter is manure or a similar high-nitrogen material, the extra fertility may not be necessary. How-ever, if the compost does not heat up within a few days after the pile is completed, then you likely need to add moisture and possibly a nitrogen source. Ideally, compost in the center of the pile will heat to a tempera-ture of approximately 170 degrees Fahrenheit. It's also important for the compost to be loose and well aerated—you need the oxygen, but you may also have to sprinkler-irrigate to encourage microorganism activity if the materials are very dry or during dry weather. So-called compost starters are of questionable value since a compost pile with the right fer-tility, water, and aeration will roar to life in a matter of days on its own. Compost piles are often built up over a period of several months or more. You will still get compost this way, but the best technique is to build the pile in a short period of time. This ensures that the pile is large enough (5 x 5 x 5 feet) to hold in the heat of decomposition and kill most weed seeds, insects, and pathogens.

Initial Fertilization

In most cases you will also want to add fertilizer to the soil. With a lot of organic matter already in the soil, the judicious use of soluble fertilizers

This small compost bin has removable front slats to facilitate removal of finished compost.

seems to make sense and poses little threat to the environment. You can get the nutrients from organic fertilizers if you choose, especially if you tend to be heavy-handed with the soluble fertilizers that can more easily burn plants. The following table lists some fertilizers—both organic and inorganic—with suggested amounts to use (per 100 square feet, per planting hole, per gallon as a soil drench). Fertilizer elements do get used up or leach out with the drainage water, and some elements like nitrogen can volatilize. For example, it's estimated that after a growing season, the garden soil loses 1 to 3 pounds of nitrogen, ½ to 1½ pounds of phosphorus, and 2 to 4 pounds of potassium per 100 square feet.

Note that these recommendations are made for level measures (¼ level teaspoon, etc.). You will probably use several materials to provide nitrogen, phosphorus, and potassium, but don't overdo it—for example, you would not add every material in the chart to a single planting hole. Also, dig the planting hole deep enough that you can cover the fertilizer with 2 inches of soil before setting the tomato transplant.

Soluble fertilizers can be used judiciously in the tomato patch. Formulations high in phosphorus, like 12-24-12, can be overdone; eventually, too much phosphorus can tie up iron and other elements. Calcium nitrate or 15-5-10 may be better alternatives if you have a heavy soil or a soil test that indicated excess phosphorus.

This new garden plot was prepared with greensand and compost. (Photo by Deborah J. Adams)

Soil Additives Used at Planting

The concept of banding fertilizer describes the placement of fertilizer under, say, a row of beans. While fertilizer banding can work for tomatoes, there's no reason to put fertilizer under the entire row when you can concentrate it under each plant.

Gardeners like to dig a deep hole or trench (if you're laying down tall, overgrown transplants); thus, the opportunity to place fertilizer/compost/mycorrhizae and other soil additives is at hand. Placing the nutrients under the root system gives the plant first access to them and keeps them farther away from the weeds. Since phosphorus does not move readily in the soil, this is one of the main elements we like to place under the tomato transplants. If phosphorus is placed 3 to 4 inches below the tender roots, the plant has a chance to become established before it gets to this "reservoir" of fertility.

One of the main microbes you might want to include in your tomato plants' subterranean cache is mycorrhizae. There are two basic types: endomycorrhizae and ectomycorrhizae. Endomycorrhizae develop in

Fertilizer recommendations

Fertilizer	Nitrogen (N)	Phosphorus (P)	Potassium (K)	Pounds per 100 sq ft for average garden soil	Amount per planting hole	Amount gallon— soil drench
Ammonium nitrate	33.5			0.5	½ tsp	½ tbsp
Ammonium sulfate	21			1	1 tsp	1 tbsp
Calcium nitrate	15.5			1.25	1 tsp	1 ½ tbsp
Chicken manure	3	5	1.5	14	¼ cup	n/a
Composted horse manure	2	1	2.5	30	¼–½ cup	n/a
Cottonseed meal	6	3	1.5	3	¼ cup	n/a
Fish meal	10	6	2	2	¼ cup	n/a
Guano	10	4	2	2	¼ cup	n/a
Mushroom compost	1	1	1	20	½ cup	n/a
Potassium nitrate	13		44	1.5	1 tsp	2 tbsp
Potassium sulfate			60	1–2	1 tsp	1 tsp
Processed cow manure	1.5	2	2.5	40	¼–½ cup	n/a
Superphosphate		20		5	1–2 tbsp	n/a, limited solubility
Urea	46			.33	¼ tsp	½ tbsp

Placing or banding fertilizer under the row is one way to concentrate the nutrients close to the tomato plants.

the interior of roots with mycelia (white, hairlike fungal strands) extending out into the soil; ectomycorrhizae develop on the exterior of roots. These beneficial organisms get food from the roots and expand the capacity of the roots to take up water and nutrients. They also form a protective barrier against pathogens. Annual plants like the tomato are primarily colonized by endomycorrhizae. You are most likely to find a blend of endomycorrhizal species sold for inoculating seeds or plants at transplanting suggested for your tomatoes. The only real test to determine if these inoculants (some include, in addition to mycorrhizae, a host of microbes like actinomycetes, threadlike bacteria that help break down organic matter into humus and help reduce root diseases and then contribute to the pleasant, earthy smell of a healthy soil) are working is to test a block without the inoculants and one with them. Be sure to use the same variety of tomato in each block (check [no inoculant] and treated [with inoculant]), and keep them separated in the garden since these beneficial organisms may migrate. Scientific studies to date have not shown a tremendous response of garden plants to mycorrhizal inoculation, but most gardeners are willing to try anything that might give them an edge. Also, one year's trial may not be sufficient to prove their benefit. Soils that are deficient in organic matter (less than 5%) are the least likely to benefit from inoculants.

Starter solutions are another way to give the transplant a head start over its weed competition. Just remember that you are dealing with a plant that is making a change in its growing environment, so a dilute solution is in order.

STARTER SOLUTIONS

Starter solutions can be made in several ways. The easiest is to use a soluble fertilizer (like 20-20-20) purchased from a garden center at one-half to one-third the recommended rate, and pour 1 pint around each transplant after setting it in the garden. You can also make your own with a granular fertilizer like 12-24-12 by dissolving 2 level tablespoons per gallon of water—again using 1 pint per plant. Organic starter solutions can be made using fish emulsion or seaweed fertilizers at the standard recommended dilution and applying 1 pint to the soil around each plant. Organic fertilizers are less likely to burn and do not need to be used at the reduced rate.

Organic Soil Additives

The following soil additives can be mixed uniformly through the garden soil with a tiller or placed under the plants in a subterranean cache. Even if you are not going 100% organic, it pays to heed some of the suggestions for soil preparation that the organic gardener uses. Rock phosphate (0-3-0) and greensand (a potassium source, 0-0-3) may not be readily available in the soil, but if you are enriching the soil with compost every time you plant, the organic acids produced will begin to make it available over time. It's like a passbook savings account for the garden—it doesn't pay much, but it gradually releases small quantities of nutrients with the help of microorganism activity and doesn't wash away with the first rain. Rock phosphate is applied at 50 pounds per 1,000 square feet, and greensand is used at 50 to 100 pounds per 1,000 square feet. They perform best if worked into the soil.

Cottonseed meal (5-2-1) is one of the most readily available and cost-effective sources of organic fertility. It helps to acidify the soil and also contributes minor nutrients. Used at a rate of 2 to 4 pounds per 100 square feet with soils that are relatively fertile and up to 10 pounds per 100 square feet in poor soils that are low in fertility, it can provide a bit of a boost while you are building up the compost in your soil. While it is pleasant to work with straight out of the bag, it does have an "old fish grease" odor if overused or when it is not mixed with the soil. Fish meal is great for the garden (it can be up 10% nitrogen), but it will get the neighbors talking—and their cats howling. A fertilizer supplier in Houston orders it for his rosarian customers, but the

Cottonseed meal is a relatively cheap organic fertilizer that is available from most feed stores.

agreement is—you pick it up the day it comes in! Adding 2 pounds per 100 square feet should be a good starting point. Blood meal, alfalfa meal, and the numerous bagged manures are also available. Try different ones and compare the production and quality of tomatoes grown with these materials to the tomatoes produced using less expensive alternatives to help you decide if these soil additives are worth the cost.

Kelp (1-0-8) is a bit expensive, but it contains a lot of micronutrients, a little nitrogen, and potassium. Some studies have shown kelp/seaweed sprays to increase the overall hardiness of plants, apparently due to potassium uptake. Roots are more extensive after seaweed/kelp treatments, and plants are more resistant to pathogens. Powdered kelp is available in 40- to 50-pound bags and can be used at 25 pounds per 1,000 square feet, or 1 pound per plant.

Gardeners close to the Texas coast should consider collecting the seaweed that often washes ashore to use in the garden. Either put it in the compost pile or use it as mulch, 2 to 4 inches deep, around your tomatoes.

Molasses is a favorite soil additive of organic gardeners. It serves as a food source for microorganisms and also contains potassium, sulfur, and micronutrients. The dry product is easy to work with if you use it right away. Once you open the bag, it soon turns into a rock from moisture absorption. The molasses available at feed stores—take your own 5-gallon jug—will work just as well for a fraction of the cost. Dilute it at 1 to 2 tablespoons per gallon and drench away. Actually, one or two drenches per year are usually sufficient. To give molasses a try, use the grocery store variety or check with nurseries that offer organic products. A quart should last a season or more.

Mulch

Mulch is typically less decomposed than compost and may be rather high in carbon. It would usually cause problems if mixed with the soil, so we use it on top of the soil or add it to a compost pile to complete the process. Pine bark is okay, and bagasse (sugar cane pulp, if you're close to the Louisiana border) is great. Alfalfa hay is excellent—it even includes some growth substances—though it's somewhat expensive, but you may need only one or two bales. Pine straw could work, and wood chips from tree-trimming crews are fine if not mixed with the soil where they will tie up nitrogen.

You have to be careful with spoiled hay. Prairie grass or coastal Bermuda may contain a herbicide. In fact, if you find one of these hays that is free of weeds, it almost surely has been sprayed with a herbicide that will leach out and affect very sensitive tomato plants. The only option for this kind of free organic matter is to pile it up and let it compost for a year or two. Then you will need to try some around a plant; in a month you should know if the plants are still being affected—if so, they will produce

Mulch is one of the best ways to control weeds, conserve moisture, and keep the soil cooler. In this instance newspaper eight to ten sheets thick was laid down (soaked in water first to prevent its blowing away), and then alfalfa hay was used to hide the paper and keep it in place.

Alfalfa hay is a bit expensive, but it breaks down quickly and contributes nutrients and growth substances to the soil.

Herbicides can leach out of grass hays like coastal Bermuda or prairie hay and damage tomato plants.

stretched-out growth and shoestringlike tips, similar to those on a virus-infected plant. Anytime you see these symptoms uniformly over a wide range of plants, it's likely to be herbicide induced. When similar symptoms show up on isolated plants, suspect a virus infection and rogue, or weed, them out immediately.

What about synthetic mulches? Weed barrier cloths can be used, but too often vigorous weeds like Bermuda grass get started in the plastic mesh and are very difficult to remove. Red plastic mulch has been shown to keep weeds down and repel insects for a while—once the foliage cover is extensive, the red plastic doesn't have much effect.

HOW TO BUILD A RAISED BED

Building a raised bed would seem to be a simple enough project. It can be as basic as using concrete blocks to outline a bed 4 feet wide and as long as you want it. If, however, you can access only one side, reduce the width to 3 feet, or you will spend a lot of time trampling through the bed trying to reach the far side. Should you decide on a wooden frame, opt for 2 x 8s to 2 x 12s. If you use boards 1 inch thick, you will be lucky if they last a year. Rocks or broken pieces of concrete make a nice bed retainer. Splash the concrete with an iron sulfate mixture at the rate of 1 cup of iron sulfate per 2 gallons of water in a sprinkler can to stain the concrete rusty red. It will look much more natural.

To slow down the grass and weeds under the bed, lay newspapers eight to twelve sheets thick before you add your soil mix. The newspapers can be laid down first before placing the wooden frame bed or setting the rocks. The paper will extend under the frame to help keep invasive weeds from creeping up the inside of the walls. An alternative technique is to lay them in after the frame is constructed and let them lap up the sides 6 to 8 inches. This will help keep a loose soil mix from washing out under the edges of the frame. Soak the papers with water first to keep them from blowing away while you place the frame or set the rocks. St. Augustine grass kills out rather easily this way, but Bermuda grass

Raised beds constructed with 2 x 12's were the beginning of the Adams Kitchen Garden. They can solve drainage and weed problems and give you a chance to custom-blend your own soil mix.

will someday rear its ugly head. If you are starting early enough that you can spray the weeds under the bed with glyphosate (Roundup, Eraser, and other generic brand names are currently available), you will get better control. This means you will have to wait two to three weeks for the herbicide to work before adding newspapers and soil. Twenty percent acetic acid would also give a quick burn down for organic growers. It's not systemic, however, so will not kill the perennial roots of weeds like Bermuda grass.

Building the beds with 2 x 12s is simplified if you use boards that are straight and free of warping, cut the boards to exact lengths, and use galvanized 90-degree corner braces—preferably on the inside. Special raised-bed connectors are available from mail-order sources (source no. 23), but they are expensive at about twenty-five dollars per set of two—

that's fifty dollars per bed. Inline connectors are also available, which enable you to add more 2 × 12s for longer beds.

Place and level the beds on flat ground, and make sure to keep the corners at 90 degrees by checking them with a carpenter's square. A scrap 1 × 2 nailed diagonally at two opposite corners will help keep beds square as you move them around. Make sure you have easy water access. You may want to dig trenches and lay some PVC pipe with risers for faucets close to the beds, or at least one at either end if the tomato patch is more than 50 feet in length.

One of the most critical factors in the success of a raised-bed garden is the soil mix. Local dirt yards (they may be listed under "sand and gravel" in the phone book) usually have a "garden soil mix" that you can have delivered. If possible, have a look at what they offer in person before you order. A mix of about one part compost, one part loam soil, and one part coarse sand or an inert material like calcined clay will get you off to a good start. It may be relatively low in fertility to begin with, but you can add more fertilizer at planting and foliar-feed, if necessary. If the mix contains a lot of undercomposted wood pieces, it will often tie up nitrogen about as fast as you can add it. Should the soil mix cost as much as or more than good-looking compost that's also available, consider using the compost and some coarse sand or "builder's sand" at a rate of two parts compost and one part sand. The salespeople at the dirt yard may even be willing to mix it up for you or, at least, alternate loading the materials in the dump truck before delivery.

Before adding the soil mix, you may want to drive several stakes in next to the inside walls and screw them to the frame to keep the bed in place. Once the beds are full of soil and plants are growing, the frames shouldn't have much tendency to move.

Gardening in raised beds will require you to be more diligent about fertilizing and watering, but you won't have to worry about the next 9-inch rainstorm drowning out your tomatoes.

GROWING TOMATOES IN CONTAINERS

Tomato gardening in containers would seem to offer a lot of advantages, but tomatoes are vigorous plants that demand a tremendous amount of water and nutrients—especially after fruit set. My best pictures of blossom end rot have all been taken of tomatoes growing in containers. Make sure to use the largest pot or tub you can live with. A 20-gallon pot is probably a minimum—a 5-gallon nursery can is a guaranteed disaster. A tomato plant in a container will need to be watered two to three times a day when it is loaded with fruit. You can set up a low-volume watering system—drip or microsprinkler—with a time clock, but a small container

Tomato plants grown in 5-gallon containers are great for getting an early start in the garden, but the containers are not large enough for production. A 20- to 30-gallon container results in a more extensive root system that can better supply water and nutrients to the developing fruit.

just doesn't have enough root volume for a mature tomato plant. When the developing fruit is stressed for water, the cells at the blossom end begin to die and the tomato turns black at the end. This is not an organism-caused disease, but it looks bad and fruit rot may set in as a result of the cell damage.

Tomatoes grown in containers are not exactly trouble free, but if you live in an apartment and the sunny patio is your only option, containers can work. Another scenario might be that the area where you would like to eventually have a garden is covered with a tenacious Bermuda grass turf that you're afraid to tackle. Maybe you just don't have anyone to build a raised bed for you—or spring is here and you don't have time to work up a garden. The excuses are many and often valid. So, yes, you can grow tomatoes in a container. Small-fruited types like cherries or the elongated cherry Juliet are excellent, but "bush" versions of standard varieties like Bush Champion are also worth a try. Just remember to water and fertilize often (slow-release encapsulated fertilizers or organic, compost-based fertilizers help keep fertility up).

Growing tomatoes in a pot can be a challenge even with large growing containers.

GROWING TOMATOES IN A GREENHOUSE

Growing tomatoes in a greenhouse sounds great in theory, but there have been many failed greenhouse tomato operations in Texas due to numerous factors, from cloudy winters to south-of-the-border competition. Those who sell greenhouse tomato systems often advertise that the plants are safe and protected in a greenhouse—you'll have freedom from

(opposite) Successfully growing tomatoes in a container means using a large container— 20 gallons or larger—and concentrating on small-fruited varieties.

pests, and in midwinter you'll be skipping merrily through the greenhouse picking bushels of vine-ripe, delicious tomatoes while hordes of buyers are lined up on your doorstep. They typically fail to mention that the pests get in the greenhouse and thrive, too, and that greenhouse tomatoes require a little help with pollination. On sunny days, you may be skipping through the greenhouse, but it will be in an effort to pollinate the flower clusters, maybe by whacking the trellis lines to somehow dislodge pollen. If cloudy weather prevails during the winter months when you want production, no technique will make heavy, wet pollen shed. Result: no fruit set. Viral diseases can be a yearly plague in the greenhouse; fortunately, we have varieties with a good deal of resistance now. A considerable amount of labor is involved—preparing the soil, seeding, transplanting, pruning, tying, picking, delivering, removing plants, and renovating for the next crop. And, except for a few high-altitude areas in Texas, summer production will be limited because of the heat.

On the plus side, there is a tremendous demand for quality tomatoes in the off season. If this type of commercial venture inspires you, then launch an Internet search for "greenhouse tomatoes." Be wary of offers that seem too good to be true—like the old 8- x 12-foot greenhouse complete with a hydroponics growing system (hydroponics is a soilless growing technique that uses a recirculating nutrient solution) and the seller will guarantee to buy all you can grow. Chances are you won't produce much, and if you do, the promoter will be long gone. If you put a pencil to the package deal, you will also find you paid a lot for comparatively little. Look long and hard before you jump into greenhouse tomatoes. Plan to attend at least one greenhouse tomato grower's short course at a major university with an agriculture curriculum like that at Texas A&M University. Aggie Horticulture is a good source of information (see the *Texas*

(opposite)
Greenhouse tomatoes grown in containers are constantly fed using a balanced nutrient solution.

Greenhouse Management Handbook at http://aggie-horticulture.tamu .edu/greenhouse/nursery/guides/green/). You can also find applicable information through Louisiana State University and Mississippi State University.

One south-central Texas grower mentioned that she was planning to sell greenhouse tomatoes nationally via eBay and the Internet. After a visit to her greenhouse, I'm encouraged that she can pull it off—and apparently the buyers really are pleading for tastier tomatoes such as Match (source no. 3) and Trust (source no. 3).

Personally, growing a half-dozen plants in the greenhouse sounds tempting, but one full of tomatoes reminds me of my college days when I lived in the Oklahoma State University greenhouse—a Quonset hut, really, but it was attached to the greenhouse. My meager, albeit free, housing came with cleanup and temperature-checking duties. Our tomatoes in the university greenhouse always seemed to be virus plagued and not very tasty. Apparently, a lot of improvements have been made in varieties and cultural techniques during the last forty years.

4 Tending the Tomato Patch

PLANTING AND PROTECTING NEW TRANSPLANTS

Always buy the healthiest plants you can find. Plants that are a bit tall can be used by laying the stem down in a trench. However, plants that look sick with a mosaic pattern to the foliage or shoestringlike tips may have been infected with a virus and are not a bargain even if they are free. Plants don't grow out of a virus, and it can spread to your other tomatoes. Medium-sized plants, 4 to 6 inches tall, are usually the best size for the garden. Tomato transplants in gallon cans are tempting if you're getting a late start, but they generally cost too much. Look for dark green stocky plants with good fertility and that have been grown with plenty of sun—in other words, not crowded. Sometimes we compromise on these parameters to get a particular variety, but never compromise by buying a sickly looking plant.

Dig the planting hole about 4 inches deeper than necessary to set the plant about 2 inches deeper than it grew in the starter pot. This gives you room to place a couple of slow-release plant fertilizer tablets in the bottom of the hole along with any beneficial microorganism stimulants—like

Greenhouse tomato transplants.

mycorrhizae. A couple of tablespoons of organic fertilizer would be even better if you're using inoculants. Almost all enthusiastic tomato growers have some special formula they like to place under the root system. Experiment with 1 or 2 tablespoons of superphosphate, encapsulated fertilizers, Epsom salts (magnesium sulfate, 1 level tablespoon in the bottom of the hole), or a handful of cottonseed meal or kelp meal under the roots. Then cover the cache with several inches of soil, set the transplant, fill in the soil, and water thoroughly. As the plant roots grow in the next few weeks, they will have this nutrient reservoir readily available. Tomato plants have the ability to send out roots from the stem, so if they are a bit tall, don't worry about planting them a few inches deep.

Remove any lower leaves that would

Setting a stocky, healthy tomato plant is easy. Dig the planting hole 3 to 4 inches deeper than necessary to set the plant 2 inches deeper than it grew in the pot. First, however, add some slow-release fertilizer to the bottom of the hole and cover it with 2 inches of soil before setting the transplant.

be covered with soil, and wrap a cardboard or aluminum foil collar around the stem that extends 1 inch below and 2 inches above the soil surface after the plant is set and the soil pulled up to the stem. If the plants are, for example, 12 inches tall, you will need to lay the stem down in a trench and carefully tip the upper 5 to 6 inches out of the trench. Pull off any leaves that will be covered by the soil, and use a small bamboo stake to support the top. Don't insist on a 90-degree bend or you may break the stem. The reason for laying the stem in a trench is to avoid losing the existing roots buried deep in a hole since they will suddenly have been thrust into a lower-oxygen environment. You can still pack some fertilizer/microbes under roots and stem, but you will just have to do a bit more digging.

Laying down a tall transplant will encourage roots to develop along the stem.

Nothing is more frustrating than to set out your precious transplants and then find them chewed off at soil level the next day. Cutworms are usually the culprit; the collar of cardboard or aluminum foil added when transplanting should eliminate the problem. Several years ago, however, tomato plants with foil collars were chewed off above the collar in my garden. No self-respecting cutworm (even the so-called climbing cutworm) is going to bridge this barrier. Remembering that field mice like to chew on foil, we came to the conclusion that our plants had been "moused." It was a minor problem, and mouse control seemed to alleviate it.

Spring frost protection is a concern that all Texas tomato growers have to be prepared for. Most years, wrapping the tomato cages with fiber row cover will get you by—especially if the frost is light. The real concern comes when temperatures are predicted to drop below 30 degrees Fahrenheit and the freeze is expected to last more than a couple of hours. If you are one of the early birds who likes to get a jump on the season

Tomato transplants can be protected from cutworm damage with a cardboard or aluminum foil collar around the stem.

by planting two weeks before the average last frost is predicted in your area, then it may be best to hold off setting cages or stakes to make it easier to cover an entire row with frost protection cloth, old blankets, or a similar material with good insulation properties. Avoid using a single layer of clear plastic. You can use one layer of plastic over the insulation material to stop the penetration of a cold wind, but don't count on it for frost protection by itself. The plastic allows heat to escape, and the plants can be colder under the plastic than they would be out in the open. A

Hard plastic covers like these can be used to cover newly planted tomatoes and provide a few degrees of frost protection.

double layer of plastic that creates a dead-air space provides good insulation, but adapting this technique to the tomato patch would be a difficult task. Small tomato plants are rather delicate, so be sure to make hoops or other support devices for the insulating cover to avoid damaging the plants.

On a plant-by-plant scale, waxed paper caps, plastic domes, and old-fashioned glass cloches are just a few of the many devices recommended to cover new tomato transplants for frost protection. Home gardeners often cut the bottoms out of gallon milk jugs and set these over the plants. They are virtually free and a good way to recycle. Leave the lid on during cold nights, and remove it during the day if the temperature is not too hot. However, if temperatures soar into the 80s, it's best to remove the milk jug entirely. Any of these individual plant covers can be used with fiber row cover cages to give the plants additional frost protection. Granted, these emergencies are a hassle, but often it only takes one or two such responses to bring your tomato crop through the last of spring's late freezes and into the lead for the earliest tomatoes on the block.

CAGES OR STAKES FOR SUPPORT

Cages solve lots of problems in the tomato patch. They keep the plants up off the ground, which helps keep the fruit away from the soil where it quickly rots and where critters find it readily. Cages support the plants for better air circulation—reducing disease problems—and make insects easier to find and picking easy. Most of the tomato cages offered at local hardware stores are flimsy and need a big heavy stake to have any hope

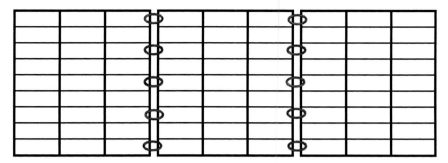

Diagram of a tomato cage with hog ring connectors.

of supporting the plant to maturity. Reinforcing wire (cut 5 feet long) makes an adequate cage, but some of the best cages in the Adams garden are made from sections of cattle or hog panels. These heavy wire panels usually come in 12-foot lengths, so you may want to take your bolt cutter along when you go shopping and cut the panels into 2-foot pieces, which will be much easier to load in a pickup or SUV with the long sides butted together. Also, buy some hog rings and a special pair of pliers to close the rings. Put three sections together, but connect only the two vertical sections in the middle. When you set the cage in the garden, it will fold into a triangle shape that is super sturdy and the other vertical edges can be tied with wire. This makes it easy to pull up the cage at season's end, untie the one edge, and store the cages flat. If you cut off the bottom rung, you can poke the vertical pieces into the ground for support, but you will also need to drive in a good stake or two on the side of prevailing summer winds (usually southwest or southeast) because a full-sized tomato plant looks pitiful blown flat on the ground by Texas winds.

Reinforcing wire cut in 5-foot lengths and wired into a cylinder makes a relatively long-lasting and stable cage provided you include a couple of 2 x 2 stakes for support. Commercial cages like those available from Texas Tomato Cages (800–983–4646 or tomatocages@yahoo.com,

Three sections of cattle panel are lined up on the vertical axis and hog rings are used to connect the middle sections.

Cattle panel sections ready to be made into tomato cages.

http://www.tomatocage.com) are expensive but a reasonable investment if you need only a few.

Regardless of the cage you use, you will also have to tie some tomato limbs—they always seem to grow out of the cage and then flop to the ground. It's almost impossible to make tomato plants grow into a perfect cylinder, so use plastic tie tape to pull the stems up and tie them to the outside. You can also weave them back into the cage and tie them on the inside, but don't get too rough or you may break the stems. We generally don't prune out the shoots from the leaf axils on caged tomatoes, but sometimes it helps to take out the growing tip of these shoots if they get too rangy—especially as the plant begins to reach the top of the cage.

The sharp edges cut by the bolt cutter can be filed or ground off to prevent snagging.

(top right)
Hog rings are used to connect the panel sections—don't compress the rings overly tight—the tips just touching are about right. Otherwise the panels may be hard to fold into a triangular shape.

(bottom right)
The extra pieces of the cattle panels can be placed in a vise and bent to make sturdy and long lasting "hold down pins" for irrigation tubing.

Newly planted tomatoes look lost in such a big cage, but they soon fill it up.

Finished tomato cage constructed with cattle panels. Note that three sections were combined using hog rings on two joints only. When placed in the garden, they form a triangle and the final joint is tied with twist ties. Two sturdy stakes or one metal T-post is used to ensure that the cage does not blow over in a high wind. At season's end the cage can be spread out and stored flat.

Once this cage fills up with foliage and fruit, two sturdy stakes will be necessary to hold it upright in a Texas windstorm.

Some folks like to stake their tomato plants rather than use cages, and you can get grand champion big tomatoes this way, but it does require more pruning, and the total yield per plant will be less. Stakes need to be substantial to hold up a big tomato plant, so plan on solid 2 × 2s as a minimum, and check out the metal bamboo-looking stakes—for extra strength combine them tepee fashion in threes for a rock-solid stake. As plants grow, you will want to snap out the shoots that grow from the leaf axils and continue tying the plants to the stakes. Also, removing 8 to 12 inches of lower leaves after the plants are about 3 feet tall (caged or staked) will improve air circulation and lessen diseases like early blight (anthracnose).

The Florida weave is a more commercial system for supporting tomato plants, but there is no reason an enterprising tomato grower couldn't use this system in a big home garden. You will need 7-foot T-posts for the ends of the row—and more if the row is really long and depending on the sturdiness of the down-the-row stakes; plan on using a sturdy one every 10 feet for long rows. The end posts are slanted out and can be further supported by deadman anchors buried or screwed into the ground. Plants are generally spaced 3 feet apart, and stakes are set between every two plants—use metal stakes if possible because the wooden or bamboo stakes will not support a heavy tomato plant in a Texas windstorm. Nylon, heavy-duty, nonstretching twine (there is actually a tomato twine—see source no. 3) strapped to your waist and threaded through an 18-inch section of 1-inch PVC—think wand with twine running through it to assist you in pulling/wrapping the twine—will, no doubt, identify you as a serious practitioner of the Florida weave tomato support system.

Start tying the twine to the end stake 6 to 8 inches above the ground (when plants are about 1 foot tall), and pull the twine around the outside of the first two plants; then loop it around the stake between them and the next two plants. You will catch the other side of the plants when you reach the end of the line and come back with the twine. Go to the other side of the next two plants, and loop the stake, and so on. Keep the line as tight as possible and tie off at the far stake. Continue doing this until you can't stand the sight of another tomato plant. If plants get tall and floppy, beg your other half or best buddy to help out—and no, you can't go on a two-week vacation! You might want to consider running rows the same direction as prevailing summer winds to avoid a direct confrontation with the wind.

An alternative solution would be to run a line of goat or hog fence

Tomatoes can be staked but requires more pruning and tying.

Removing the lower tomato leaves when plants are knee high helps improve air circulation and reduces the incidence of disease.

The Florida weave support system for growing tomatoes is adapted to large-scale growing operations.

Twine with little or no stretch is used to weave the planting row into a hedge of delicious red tomatoes.

with 4-inch or larger square holes supported solidly with T-posts, and weave the tips of plants through the fence as they grow. This should give good support, but getting the dead plants out of this fence at season's end might be a nightmare.

Unfortunately, letting the plants sprawl on the ground just isn't an option. Diseases from leaf spots to fruit rots will be rampant, and critters from rodents to the family dog will enjoy the fruit before you do.

THE ROW-COVER BENEFIT

Fiber row cover is a relatively new addition to the gardener's arsenal of climate and pest-control options. It's basically diaper-liner technology— a spun polypropylene material—that is used as a tunnel cover over a row of veggies or, in the case of tomatoes, usually wrapped around the tomato cage in early spring to protect the plants from drying winds. It also gives a few degrees of frost protection and may serve to keep pests out, although the bugs really thrive under the cover if they were already present when the plants were covered. In defense of the "nothing new

Tomato cages with fiber row cover.

Tomato plants are cozy and safe from drying winds, insect pests, and light frosts in a cage wrapped in fiber row cover.

under the sun" theory, the glass cloche served a similar purpose more than a century ago, but making one large enough to cover a tomato cage would be unwieldy at best. To some extent, using row cover makes the technique of "hardening off"—allowing the transplants to stay a bit dry and colder—unnecessary, since with row cover the plants hardly know they have left the comfort of the greenhouse or cold frame.

After setting out tomatoes and putting the cages around them, wrap row cover around the cage with extra at the top to tie off if a frost threatens,

Glass cloche, a small, portable plant covering designed to be easily moved around the garden traps the sun's warmth raising the temperature of both the air and soil inside. Fiber row cover or fiber frost protection cloth is less expensive and capable of covering large plants or entire rows.

and clip the cover to the cage with clothespins. After the plants are knee high, remove the row cover and store it for next year. Row covers 8 to 10 feet wide and cut long enough to wrap around and lap a bit are perfect for most cages. You will have extra at the top to tie off for frost protection, or open it up and roll the fabric down if you need to check for insects, spray, or foliar-feed. Actually, it's quite easy to foliar-feed through the row cover.

MAINTAINING SOIL FERTILITY AND FOLIAR FEEDING

Tomato plants that are yellow and puny are either diseased or hungry. Usually they are hungry—they need more fertility. You can use organic fertilizers—they are a little slower to get into the plant but less likely to burn the roots—or you can use soluble fertilizers added to the soil or applied to the leaves. Tomatoes demand good soil and high fertility—it's not an option!

A fairly standard recommendation used to be not to fertilize tomato plants (after the initial starter solution) until the first cluster of tomatoes is the size of a dime. This is still good advice for standard and heirloom varieties, most of which are indeterminate monsters! Especially with too much nitrogen, it is rather easy to end up with an 8-foot tomato plant and only a few fruits. This plant may be the pride of the neighborhood, but in your heart you really wanted tomatoes. Hybrid and determinate varieties require more fertilizer—they have been selected for good fruit-setting genes.

When tomato plants look spindly, light green, or even yellow, they likely need higher fertility. New organic gardeners often underestimate the amount of compost and organic fertilizer they will need.

If your plants are getting plenty of sun but look pale green and spindly, they likely need more fertilizer. You can side-dress with a fertilizer like 15-5-10 at 1 pound per 100-foot row ("a pint is a pound the world around") or foliar-feed with a hose-on sprayer. Organic gardeners have an ever-increasing array of fertilizers to use, from bat guano to domestic manures. These are often blended with greensand, rock phosphate, kelp, molasses, and other ingredients to add additional nutrients and beneficial microbes. Be sure to check directions for applying these products, and follow them to the letter. Too many gardeners get into trouble presuming that if a little bit is good, a lot will be better.

A soil test is the best way to determine how much fertilizer you need, but most gardeners don't get the testing done every year. Every other year is usually sufficient. Since the test for nitrogen is not always valid (many testing facilities base their recommendations on the questions you answer about your fertilizer regimen), you will just have to learn to recognize optimum growth based on plant characteristics. This gets easier as your garden matures and you have worked with the soil for a while. Tomatoes with stems thicker than your thumb are typically getting too much nitrogen. If you see some of the nutrient-deficiency symptoms featured in your old soils textbook (you can find these photos on the Internet, too), like purple stems indicating a phosphorus deficiency, then you likely have a way to go. Most of these symptoms don't show up in normal soils, but they may occur in container soils or in raised beds with mostly sand for a soil base.

The Texas A&M University Soil, Water & Forage Testing Laboratory (2474 TAMU, College Station, TX 77843–2474, http://soiltesting.tamu .edu/) is the place to go for your basic soil test. The lab can also test irrigation water and look for an overabundance of some elements, like boron, that may be toxic in excess. Supplement this initial test with home test kits using color charts to keep on the right track through the growing season.

Explaining N-P-K

N stands for nitrogen; P, for phosphorus; and K, for potassium. These three elements are required by law to be listed on fertilizer bags. Often the percentages of other elements like sulfur (S) and iron (Fe) may also be listed by the formulator. It's an oversimplification, but nitrogen does tend to encourage foliage growth, phosphorus is important for flower and fruit development, and potassium is vital in cell chemistry and the overall health and vigor of the plant. Any one of these elements can be limiting. If your soil has plenty of nitrogen but is phosphorus deficient, the plants are not going to make the rapid vegetative growth that high-nitrogen nutrition is noted for. Before you can get the benefit of the nitrogen, you would have to add a soluble form of phosphorus, for example, superphosphate (0-20-0) or a fertilizer like ammonium phosphate (mono = 11-52-0, di = 18–46-0) that is high is phosphorus. Fortunately, your soil test results from the soils lab at Texas A&M University will include fertilizer recommendations, so this does not have to be as complicated as it may sound.

Minor nutrients like zinc, magnesium, calcium, molybdenum, and boron can also be limiting or even harmful in excess. They may or may not be found in a standard bag of fertilizer, but they are often available in organic fertilizers and can be added from specialty fertilizers like iron chelate, zinc chelate, or a micronutrient blend. Chelates are formulations that are resistant to tie-up—usually because of high pH soils or irrigation water.

Soil pH, a Key to Success

The ideal soil pH for tomatoes is 6.5 to 6.7, or slightly acidic. Soils that are very acidic, as might be expected in East Texas, may require lime—calcium and/or magnesium sulfate. It's best to apply lime based on a soil test, but if you have a sandy soil in a high-rainfall area and the garden plot has never been limed, spread 5 pounds per 100 square feet mixed thoroughly into the top 12 inches. Heavy clay soils and most other soils in Central/West Texas typically have a high or basic pH in the range of 7.2 to 7.8. Soils with a pH above 8.0 are marginal for tomatoes though suitable for creosote bush. Fortunately, you can amend almost any soil, and organic matter is the key. Organic matter helps sandy, acidic soils hold moisture and nutrients. Heavy alkaline soils have improved soil structure (granularity) with the addition of compost, which also helps hold nutrients, encourages beneficial microorganisms, and improves drainage. Again, the gardener with the most compost wins!

Quick-Start Fertility

Here's the all-too-familiar scenario: you have a garden plot dug up, you just bought plants at the nursery, and you're ready to plant. You can't wait weeks for a soil test to come back. Generally, 1 to 2 pounds of a complete fertilizer like 15-5-10 per 100 square feet of bed area is a good place to start. Why not use 13-13-13 or 12-24-12? Phosphorus, the middle number, tends to accumulate and can tie up other nutrients like iron. An organic fertilizer with nitrogen, phosphorus, and potassium (in that order) can also be used—just follow the directions on the individual product label.

Foliar Fertilizer

There are also benefits from foliar application of fertilizers and liquid compost (compost tea, etc.). Rates of application vary based on the product, but typical for a soluble fertilizer like 20-20-20 is 1 tablespoon per 1 to 2 gallons of spray water. Use the more dilute solution in hot weather and the concentrated rate in early spring when plants are making active growth.

One of the easiest ways to get a boost from foliar fertilizer is to use a hose-on fertilizer applicator. These devices are designed to meter fertilizer from the saturated solution at the top of the jar. Thus, you pour a package of concentrated fertilizer into the container, fill it with water, and spray away. Hose-on sprayers with a tube into the jar have to be set to an amount, such as 1 tablespoon per gallon, and they are more appropriate to application of pesticide sprays.

Compost tea is one of the organic gardener's versions of foliar feeding and may also contribute other plant health benefits that are just being discovered, like promoting colonies of plant-healthy microbes on the leaf surface. Consider yourself lucky if you live in a community where there are nurseries specializing in organic products, as you can let them do the brewing for you. Compost tea advocates generally recommend applying the product within 24 hours of bottling since the beneficial organisms die off rapidly without aeration. Even though compost tea contains a lot

Foliar feeding can give a big boost to tomatoes; however it should be used as a supplement to good soil nutrition.

of good "bugs" beneficial for plants, don't be too cavalier about using it. Wash your hands and arms after applying it, and keep away from any spray drift if you use a sprayer to put it out. If you must make your own, there are compost tea systems or units that keep the material aerobic and safe to use. If not aerated properly, anaerobic, hazardous bacteria may be produced. The Pennsylvania Department of Environmental Protection has a Web site featuring a homemade unit made with a 5-gallon bucket and aquarium supplies for the do-it-yourself gardener (http://www.dep.state .pa.us/dep/deputate/airwaste/wm/recycle/Tea/tea1.htm). How often do you need to use the tea? Some gardeners get by with one to two sprays a season; during times of stress or disease pressure you might use it weekly or biweekly.

Organic fertilizers like fish emulsion or liquid seaweed are easier-to-find alternatives for the organic gardener to use for foliar fertilization. They are less likely to burn than soluble fertilizers, so they should work fine full strength, but in hot weather it would be prudent to use them at half strength on the foliage during the cooler morning and evening hours, or at least test the mixture on one plant before treating the entire garden. They may plug up hose-on sprayers, so you may be more successful applying them with a watering can.

IRRIGATION—NOT AN OPTION

Texas weather is often a matter of alternating floods and droughts. It would be nice if we could space out the occasional 9-inch rain, but short of collecting water in tanks and saving it for irrigation, we are at the mercy of the weather. Actually, water collection is a hot topic these days, and all Texans would be wise to check it out (see the Aggie Horticulture Web site: http://aggie-horticulture.tamu.edu). In the meantime, we need to plan for the efficient use of water with low-volume irrigation systems. While furrow irrigation will certainly work—just fill up the garden furrows and let the water percolate into the soil—it is rather wasteful. It may be a useful technique after a really dry spell or as a way to increase soil moisture at the start of the garden season, but generally we will be using drip, microsprinkler, or bi-wall watering systems.

These low-volume systems are easy to install, and most garden centers and hardware stores have one or more types available. A wealth of information and irrigation materials is available on the Internet. The tools are simple, but it's best to start with a plan. Virtually every manufacturer includes sample plans to use as a "go by" that you can adapt to your garden.

You will need a filter, backflow preventer, pressure regulator, pocketknife (scissors-type pruners work well, too), connectors (Ts, elbows, in-line connectors), poly pipe (½ or ¾ inch plus spaghetti tubing if you want to move drippers or microsprayers away from the main line), a punch for

Collecting rainwater will be increasingly important in future gardens.

installing certain fittings and drippers into the poly pipe main lines and microsprayers, and drippers or bi-wall tape—which one you use depends on the crop to some extent. Hint: Buy the best punch you can find, and make sure it is designed for the fittings you plan to use. Don't forget to purchase a package of plugs to fill the holes punched into the poly lines when you decide to move a dripper or you make a mistake and punch a hole in the wrong place.

Your irrigation system will require a water hookup—usually a faucet, but you can have a separate valve just for the irrigation lines. Between the water source and the main line (also called the header line), you will want to include a filter, a backflow preventer (to prevent backup of dirty water or fertilizer water into the house water line), and a pressure regulator (city and well water pressures are too high). Then you will run a

A

B

C

D

E

F

Drip irrigation: A) a bubbler is a dripper that puts out more volume but not so much that it causes flooding; B) half-inch poly header line with drip tape (bi-wall) attached; C) three drip tapes (bi-wall) in a raised bed; D) bi-wall or drip tape pushed onto the fitting and secured with a twist fitting; E) drip tape leaks water through slits at intervals along its length (Photo by Deborah J. Adams); and F) a typical garden dripper is less likely to plug with silt when turned up but is photographed turned down to show the drip action.

Irrigation fittings include connectors, backflow preventers, filters, pressure regulators, drippers, microsprayers, and drip tape.

½- or ¾-inch poly line along the edge of the garden at a right angle to the tomato rows (see irrigation plan).

Unless you have a hose faucet close to the garden, your first challenge will be to run an underground PVC line to the garden and bring up a few pipes with faucets. Note: Instead of a poly main line, you can run an underground PVC line for the main header with risers, elbows, and a connector at each row for attaching the ½- or ¾-inch poly pipe that will run down the rows with drippers (the connector to go from PVC to poly has a compression fitting for the poly connection). An alternative would be to attach bi-wall tape directly to the PVC connector for each row with the appropriate fittings. Just remember that it is more work to move the drip lines with PVC header lines since they are rigid, and you can't just fill the holes with a plug like you can with poly.

Since we are concentrating on tomatoes, the drippers and bi-wall tubing (drip tape is a synonym) work best because the microsprayers tend

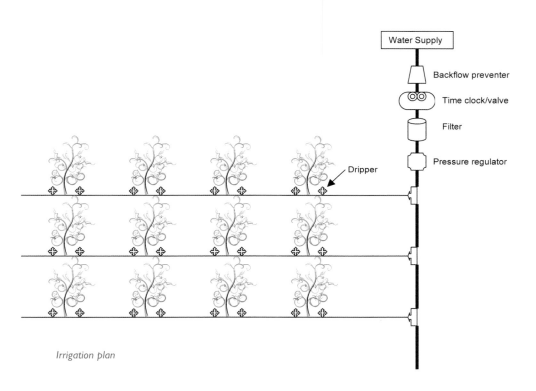

Irrigation plan

to wet the lower foliage and encourage disease infection. Drippers come in many styles—some drip and some bubble—the latter with higher volume but not enough to cause erosion. Always try to use the same drippers throughout the system or you will end up with wet spots and dry spots. One or two drippers per plant—depending on the volume emitted (usually 1 or 2 gallons per hour)—should keep the plants thriving all through the spring and into the summer. In most cases it's best to locate the drippers between plants rather than right at the base. This tends to give a more even distribution to the feeder roots that develop closer to the foliage drip line than near the stem.

The bi-wall tubing or T-Tape comes in a roll that you lay out along the row and cut off as needed. There is a top side to this material that should be printed on the tape. If you put the leaky side down (it has small slits at intervals, usually 8 to 12 inches), the line tends to clog up faster. The punch-in connectors often sold with the bi-wall tubing that come out of poly main lines have a tendency to leak at the connection, so if you can find T-connectors that fit in the line, you'll likely be happier. You may also find connectors with a valve that will allow you to shut off a line that is not needed. To close off the end of the lines, precut a short piece of T-Tape (1½ to 2 inches), then simply fold the end of the row tape over a couple of times, pinch along the sides, and slip the precut T-Tape over the folded end.

Hook up a battery-operated clock/valve, set it for 30 to 60 minutes per day, and you can go on vacation and still come home to a thriving garden. If you do leave town, you should go over the system with someone who can drop by once or twice a week just to make sure it's working properly. If you want to go really high tech, you could add a fertilizer injector to the system and just come back in June to harvest the crop. Not really, of course; everything needs some maintenance, but the fertilizer injector could help you maximize production.

CLEANING UP FOR THE NEXT CROP

Texans are fortunate to have a long growing season, but the midsummer temperatures and humidity can bring a halt to the tomato harvest. If at this point the leaves are covered with spider mites or most of the leaves have fallen due to early blight, it might be best to pack it in and dream of next year's harvest. Fall tomatoes are not always successful; if you must try, it is usually best to replant with cherry tomatoes or other small-fruited types.

Wading into a garden full of overgrown dead and dying tomato plants isn't most folks' idea of "big fun." In fact, it is especially miserable in the heat of summer or early fall. But it has to be done to get ready for fall crops or the next tomato crop in the spring. How do you extract an 8-foot tomato plant from a 6-foot cage with stakes and ties and dead and live stems and foliage sticking out everywhere? Start with a wheelbarrow. Load it with pruners, maybe even a lopper, a good pair of gloves, and a hat. Wear an old long-sleeved shirt to keep your arms from turning green. Throw on overalls or old jeans and old shoes or boots. Include a small ice chest with water or whatever beverage it takes to keep you at the job until it is done. Unload the wheelbarrow near the first plants, take a long cold drink, put on the gloves, pick up the pruners, and start cutting tomato stems. Watch out for wire—it's hard to cut. Be sure to remove all the plastic ties you used to train the plants in the cage. Pull the stems out—up or down depending on the path of least resistance. Plan to cut branching stems where they make a Y that catches the cage as you pull. As much as possible, cut rather than pull. Things soon begin to sort out faster than you thought when you first considered the project. If the plant is not too diseased or full of mites, haul the plant material to the compost pile or to the burn pile if that's an option. Rake up as much of the old tomato leaves and other debris as possible, and build up the beds with more compost and fertilizer to get ready for the next crop. If you have cages that will collapse for storage, put them up for next year and get on with the fall crop—broccoli, cauliflower, lettuce, onions, and radishes thrive during the cool season in most Texas gardens.

In some areas of Texas—mainly in the High Plains and North Texas—tomatoes may come back for a good crop in the early fall, but in south-central Texas where we garden, they are usually a disaster. They may show some promise in late summer, but fall rains and heat in September typically bring on disease and reduced fruit set—blossom drop is common when temperatures are in the 90s. Even though it's hard to pull out a plant that appears to be making a recovery, most years it is best to clean house and get ready for a great cool-season garden that makes our Texas gardens a year-round event.

We would all like to have enough room to rotate crops so we don't have to plant tomatoes and close relatives in the same spot each year.

Cutting stems inside tomato cage.

Cutting ties inside tomato cage.

Pulling out the plant after cutting stems and ties.

The problem is that tomatoes have lots of relatives—peppers, eggplants, tomatillos, and potatoes. This makes such crop rotation virtually impossible. We just have to do what we can—plant the tomato patch in sweet corn every few years, or let it lie fallow, or plant that section of the garden with a green manure crop like southern peas for a season. Sometimes planting disease-resistant varieties and adding more compost and fertilizer are the best we can do.

(opposite)
This is why you don't want to save tomato plants for fall production.

5 Tomato Varieties

The numbers in parentheses in the following lists refer to sources in the Tomato Source List near the end of the book.

HYBRID VARIETIES—MEDIUM TO LARGE FRUIT

This list usually begins and sometimes ends with the 1984 All-America Selection (AAS) variety Celebrity. The author is not a big fan of Celebrity at the table, but it sure beats most grocery store tomatoes hands down, and if you love Celebrity, that's fine. Everyone is looking for something a little different in a favorite tomato. Beyond this most famous of hybrid tomatoes, there is a bewildering array of established varieties and a yearly plethora of new ones to try. It's a bit frustrating, and some gardeners find one they like that's also easy to grow with good pest resistance, so they settle on perfecting their tomato crop with one or, at most, two or three varieties. That's a sure formula for success, but like Sirens in the garden, the new and old varieties beckon to most of us, and we are continually experimenting while keeping a core of favorites in the garden each year.

Thirty-one years as a county extension agent with volunteers who tested different varieties in their gardens, including twenty years with a Harris County demonstration garden and an army of Master Gardeners to make it work, has given the author a special opportunity to evaluate a number of tomato varieties. Since my retirement in 2001 and move to a rural area (between Houston and Austin), we have developed a large kitchen garden with room for far more tomatoes than we can consume. We sell a few at farmers' markets, but mostly we share them with our friends. The following list highlights some of the varieties grown and consumed at the Adams household. Varieties listed as "for trial" haven't been tested thoroughly, or they just sound good in the catalogs. Tomatoes not billed as overly large and noted as being tasty and disease resistant get the most consideration for a place in our tomato patch. Home gardeners should expect a minimum of 30 to 40 pounds of fruit per plant, with the overachievers harvesting more than 50 pounds per plant. Eat all you can, sell some, give some away—can and freeze the rest. Tomatoes taste best at room temperature or even at field temperature—that's why they seem so flavorful munched right out of the garden. If you must put them in the refrigerator, take them out two hours before serving to enhance the flavor.

Tomato varieties are generally described as determinate or indeterminate. The indeterminate varieties continue growing until they are big enough to cover a small house. Unless stated otherwise, the varieties in this list are considered indeterminate. The determinate varieties (sometimes referred to as "bush" varieties) are supposed to stay under 36 inches; the semi-determinate varieties grow a bit taller (3 to 5 feet in theory), but they won't take your house. In Texas gardens, semi-determinate tomatoes often grow out of a 6-foot cage and thus demand enough respect that at least two stakes or one metal T-post are needed to keep the cages upright with a fruit-laden plant in a windstorm. Finally, dwarf varieties can be grown in hanging baskets or small containers. They are cute, but don't stock up on canning jars.

There are a lot of varieties, old and new, to try.

(opposite) Probably the most popular hybrid tomato in Texas, Celebrity is a heavy producer with good disease resistance.

Harris County Master Gardeners conduct a blind taste test of tomato varieties

Unfortunately, tomato varieties come and go. Two French hybrids are especially missed of late—Dona and Carmello. Both of these Villmorin hybrids had excellent disease resistance and superb quality. Dona was the smaller of the two—4 to 6 ounces—with a typical flattened tomato shape. It was a heavy producer of delicious tomatoes. Carmello was equally good with more of a globe shape, and it was often twice the size of Dona. Carmello still shows up in a few catalogs, but one wonders if it is a true hybrid. Apparently these varieties were so popular that a lot of gardeners and growers saved their own seed since the seeds were not available in hybrid form from the parent seed company.

Hybrids are created by crossing two separate breeding lines—one of which may contribute disease resistance, and the other, quality. This is an oversimplification, but the hybrid vigor comes from crossing two very different parents to produce one uniform and vigorous offspring generation. If you save seed from this hybrid, the genes begin to combine differently so that you may have three good plants, four strong growers with mediocre fruit quality, and three plants only a mother tomato could love. These plants are referred to as an F2 generation; the next saved seed plants are an F3, and so on. The percentage of good ones tends to decrease with succeeding generations. Eventually, however, a fairly good open-pollinated variety could be selected from these offspring by roguing out the weaklings and propagating the plants with the best fruit and disease resistance. Hopefully, these hybrids will become available again someday.

DISEASE-RESISTANCE ABBREVIATIONS

You will notice a number of letters after the names of most of the hybrid varieties. These refer to the types of disease resistance that these varieties carry. It does not mean that they are immune to the disease, just resistant. Picking a variety with a large alphabet seems like the way to go if you don't want to spray or if a viral disease is prevalent in your area, but no one has yet to code flavor or texture to the variety. Maybe there should be a Flv or Txt designation for flavor and texture. I'm not sure we could trust the seed purveyors though since I can't recall seeing anything but praises for flavor and quality. If you do an Internet search, you will often find a quality rating included with university variety trials. You are still relying on someone else's taste buds, but the information is at least helpful. Here are some of the tomato disease-resistance codes to look for. Not all of these diseases will be prevalent in every Texas garden.

V—verticillium wilt

F—fusarium wilt, FF—races 1 and 2, FFF—races 1, 2, and 3 (races are different strains of the same organism)

N—nematodes

T—tobacco mosaic virus

A—alternaria stem canker

St—stemphylium gray leaf spot

TSWV—tomato spotted wilt virus

TYLC—tomato yellow leaf curl

Note: V usually means "resistance to verticillium wilt," but when combined in a series of letters like TMV or TSWV, it means a virus. In an attempt to eliminate some confusion, single-letter abbreviations are included without a space (N = nematode, F = fusarium, V = verticillium), and multiple letters for resistance to a single disease (usually a virus) are separated by a space. It should be noted that a more complex system has been proposed by the scientific community, but, to date, it has not been widely used in seed catalogs.

Amelia Hybrid VFNSt TSWV (21, 22)—75 days—was developed for southeastern U.S. growers and is notable for resistance to tomato spotted wilt. Plants are determinate to semi-determinate with loads of 7- to 8-ounce fruits. The fruits are also crack resistant and pretty decent at the table, especially when vine ripe. Where spotted wilt is a problem, this could be your main variety, though it's not quite luscious.

Better Boy Hybrid VFNASt (2, 8, 16, 19, 20, 21, 22)—75 days—has been around long enough to start applying for heirloom status, but it is still a reliable producer of medium-large to large fruits with excellent quality. This is one vigorous bush with plenty of foliage to protect tender fruit from sunscald in a Texas summer. In fact, at times, you may have to hunt for the tomatoes. It started as a replacement for the equally delicious Big Boy Hybrid, but unfortunately Big Boy can be a little light on production.

Better Bush Hybrid VFN (7, 9, 20)—68 days—was one of the first tasty determinate varieties to be trialed at the Harris County extension gardens. Now the tendency seems to be producing bush varieties of favorite standard varieties—for example, Bush Champion. This one, however, remains a valid option with a good sweet tomato taste.

Bush Champion VFFASt (19, 20)—65 days—is similar to its taller cousin Champion and worth a try if you have small cages or want to grow it in a container.

Bush Early Girl VFFNT (7, 9, 20)—67 days—is a very compact plant with the quality in an early tomato that you expect from midseason varieties. Given good soil, fertility, and water, it will continue to set in hot weather.

Amelia is a productive hybrid on a compact bush.

Bush Goliath Hybrid VFN (20)—68 days—looks like an heirloom, quacks like an heirloom, but has modern disease resistance. It has good-quality firm fruit—part of an extensive family of Goliaths that ranges from Italian Goliath VFFNTA to Sunny Goliath VFN. These varieties are exclusive to the Totally Tomatoes catalog.

Bush Champion is similar to Champion Hybrid on a more compact bush.
Fruit quality and productivity are not quite as good.

Carmello Hybrid VFNT (9, 10)—75 days—is a large-fruited French hybrid variety with scrumptious fruits. Along with Dona, this variety has not been as available as in the past, but its loss in the tomato patch would be a tragedy. Some sources may be offering seed and plants from saved seed, and these nonhybrids (or F2, F3 versions) should produce good fruit but without the hybrid vigor.

(opposite) Carmello, a French hybrid, has been a favorite large tomato of
Southern growers for years now but may not currently be available as an F1 hybrid.

Celebrity

(opposite)

Champion is a great-tasting, relatively large tomato that has been a mainstay in the Adams garden for years.

Celebrity Hybrid VFFNTASt (widely available as seeds and transplants)—70 days—is likely the most-planted home garden variety around. It's good, a lot better than grocery store tomatoes, but not one of the author's top favorites—a bit firm and not as zingy in flavor compared to Champion. It's considered a semi-determinate variety, and a Bush Celebrity variety has been offered in past years.

Champion Hybrid VFNT (19, 20)—62 days—is a mainstay favorite in the author's garden. Good disease resistance and abundant, very tasty fruits make this one hard to beat. Like all tomatoes, the fruit size slips in the summer heat, and cracks can be a problem, but you'll be tomato saturated by the time this happens.

Early Girl Hybrid VFF (widely available as seeds and transplants)—57 days—has nice 4- to 6-ounce, slightly flattened fruits that are very tasty for an early-season variety. The texture is firm with no graininess. This is a longtime favorite destined to someday make its way into the heirloom category.

Early Goliath VFFNTASt (20)—58 days—produces large 8-ounce fruit with that sweet heirloom quality in a modern hybrid. This popular early member of the Goliath family from the Totally Tomatoes catalog has been a good producer of luscious tomatoes without the grainy texture of some large

tomatoes. Others include Bush Goliath Hybrid VFN—68 days—a determinate version with high sugar content in 4-inch fruits; Cluster Goliath Hybrid VFFT—65 days—with 4- to 5-ounce fruits in clusters; the original Goliath Hybrid VFFNTASt—65 days—and larger fruit; Italian Goliath Hybrid VFFNTA—76 days—beefsteak type with green shoulders; Old-Fashioned Goliath Hybrid VF—78 days—large, flattened with rough shoulders; and Sunny Goliath Hybrid VFN—70 days—a mild, low-acid, yellow/gold version with 7- to 8-ounce fruits.

Fantastic Hybrid VF (16, 17, 19, 20)—65 days—is another survivor of the early days of hybrid introductions. It is still a good medium-large, firm, and tasty tomato with resistance to cracking and gives heavy production from a large, indeterminate plant.

Lemon Boy Hybrid VFNASt (16, 19, 20, 22)—72 days—is a high-yielding yellow tomato with mild flavor and 6- to 7-ounce fruits. If you are looking for a little color variety without giving up productivity, this is the one to plant.

Margo Hybrid VFT (6, 20, 22)—70 days—makes for an extremely productive, rather compact determinate tomato with virtually perfect 5- to 6-ounce fruits. Quality is good—maybe not scrumptious but much better than grocery store tomatoes—and you will have an abundance to share or sell.

Odoriko Hybrid (4)—75 days—is one of the few pink tomatoes that really has flavor. The tomatoes are medium sized, and the plant is a vigorous indeterminate. Momotaro Hybrid is another pink one with greenish shoulders that has sweet flesh and a tough, crack-resistant skin. Both of these Japanese varieties are popular in Asia and grow well in Texas.

Early Goliath is a good example of an heirloom-like tomato with current breeding for flavor and pest resistance.

Fantastic Hybrid has been around for a while, and there's even a Super Fantastic. Both are still good varieties.

Two very different "Boys," Better Boy and Lemon Boy are both productive and tasty.

Odoriko does not sound very appetizing, but it is one of the best of the Japanese pink varieties.

Tomande Hybrid VFFNT (19, 20, 22)—68 days—is an Italian-style slicer. The shoulders are ribbed and can be a bit green, with green core tissue at the shoulder that needs to be discarded, but the rest of the tomato is worth the effort. This tomato is sweet, flavorful, and large enough in the early season to lap over a burger. Like all varieties, the size dwindles with the heat of summer.

Tomande, an Italian-style tomato, is juicy and delicious with just the right texture— firm but not hard or grainy.

Many of the large tomatoes like Tomande have an extensive green core at the stem end. Fortunately, the rest of the tomato is worth it.

Top Gun Hybrid VFA TSWV St (21)—75 days—is a heavy producer of high-quality, that is, tasty with good texture, 8-ounce tomatoes on a determinate plant. In our Texas climate it will still grow to the top of a 5-foot tomato cage, but it's manageable. This variety also has resistance to tomato spotted wilt virus, which is common in Texas. Top Gun is a variety to try even if viral diseases have not been a problem in your garden.

Tycoon Hybrid TYLC V TSWV F(1, 2) VNT (not currently available as a retail variety)—75 days—is a large oblate (round and flattened) 10- to 12-ounce tomato with extensive disease resistance from Hazera Seed Company. It has almost the entire alphabet after its name. Apparently it is popular with "green wrap" growers because of its uniformity, size, firmness, and quality. Unlike most home varieties, the stem stays on the plant when the tomato is picked. This makes the fruit easier to ship since there's no stem to poke its neighbors and start the decay process. The plants are considered determinate, but they will fill a 5-foot tomato cage. It also has good heat-setting characteristics.

For trial: Bella Rosa Hybrid (10, 19, 21), Momotaro (4, 6, 17, 19), Muriel Hybrid (21), Solar Fire (2, 16, 19, 20), Solar Set (19, 20), Sun Leaper (19, 20), Sunmaster (19, 20, 21, 22), Talladega Hybrid (19, 20), Ultra Sweet (16)

OPEN-POLLINATED AND HEIRLOOM VARIETIES— MEDIUM TO LARGE FRUIT

These varieties typically have little or no disease resistance, and some of the heirlooms are downright ugly with off colors, pleated shoulders, and grotesque shapes. Usually these physical detractions are no measure of the quality and flavor that these varieties bring to the table. The legendary Brandywine is out of reach for most Texas gardeners—it just doesn't

(opposite) Tycoon Hybrid is not readily available to the home garden market, as it was developed for commercial "green wrap" production, but it is very productive and tasty when harvested vine ripe.

Brandywine is perhaps the most famous of the heirloom varieties and also one of the most difficult for Texas gardeners to grow.

set fruit in hot weather, and insects and diseases love it. If you "just gotta grow one," try the grafting technique described earlier to, at least, provide a vigorous, nematode-resistant rootstock. If you want to offer heirloom varieties at a farmers' market, try packaging them as an assortment in a bag or sealed tray. Buyers will then be forced to try some really good tomatoes that they would not have picked out of a flat if left to choose for themselves. Having some samples cut up and skewered with toothpicks for taste testing is a great idea, too. Be sure to keep them covered so the flies don't spoil the moment, and have a shaker of salt handy for folks who want to try them that way—salt is a flavor enhancer and may help to seal the deal.

Heirloom tomatoes in the marketplace do not have the uniformity of commercial varieties, but the flavor can be outstanding.

Standard varieties don't have hybrid vigor, but they are usually quite uniform from years of selection and should breed true from saved seed. These varieties include Homestead, Rutgers, and Marglobe, which made up the majority of varieties planted in the gardens of our parents and grandparents.

Black Krim (1, 11, 13, 15, 19, 20)—69 days—is one of the Russian black varieties. It's early with blackish, deep red fruits that are delicious. Fruit tops tend to be a bit green and rough, and there is almost always some loss of quality at the stem end on these heirloom varieties. If you like to experiment with heirlooms, this is one to include.

Cherokee Purple (1, 6, 11, 13, 15, 17, 19, 20)—80 days—is a large rose-purple heirloom that has grown well in Texas. The fruits are somewhat

This Russian heirloom, Black Krim, is an early-fruiting example of the popular black tomato. Who could resist a burger or BLT with slices of this exotic Black Krim tomato?

perishable, and the shoulders are a bit green with a coarse green core at the stem end, but the rest is tender with a delicious, complex flavor. Cherokee Chocolate (1, 19) is reported to be an improved form.

Georgia Streak (15)—90 days—looks like a tomato you might get on your fast-food burger in January, but it tastes great and produces fairly well in the Texas heat. The fruits are large—often over a pound—and yellow streaked with red.

German Johnson (1, 14, 15, 19, 20)—80 days—is a large-fruited heirloom that has produced well in south-central Texas. Considering the heat, humidity, and pests in this area, this one should qualify for trial in most areas of Texas. Fruits can weigh up to a pound and are meaty and tasty.

German Johnson is a large-fruited heirloom with good flavor and is a good producer in Texas gardens.

Golden Jubilee (1, 2, 19, 20, 22)—72 days—is a medium-large orange tomato with mild flavor and low acidity. This variety is not really the author's "cup of tea," but if you're looking for color and not much zing, this one is a vigorous plant and a good producer even though it lacks pest resistance. It should be a good one for farmers' market sales since some people like a mild tomato.

Green Zebra (1, 3, 6, 10, 11, 13, 14, 17, 19, 20)—75 days—is a plant breeder's nightmare, at least if the plant breeder works for your average seed company. Tom Wagner is anything but average. He is an amateur breeder from California who introduced this tomato long ago without any patent

Often listed as an heirloom, the Green Zebra variety is a relatively modern creation of Tom Wagner that looks odd enough to blend in with the "heirloom crowd." It's a refreshing addition to any salad.

protection, and it caught on. It's not even close to being an heirloom, but it looks like a loser (loser and heirloom are synonymous in many cases), so it got swept along with the rest. He also breeds potatoes in various colors with less notoriety. In spite of its looks, this is not a bad-tasting tomato, especially when tinged with a bit of yellow-orange. Most would describe it as refreshing.

Homestead FA (19, 20, 22)—80 days—is an older commercial variety for Southern producers. It might not quite hold its own with modern hybrids, but it produces a quality 8-ounce tomato on a plant that is somewhat determinate in growth habit.

Marglobe, Marglobe Select, Marglobe Improved VFA (15, 19, 20)—72 days—are upgrades on an old favorite. Grandma's garden always had a few Marglobe and Rutgers plants—the new guys in her garden were Big Boy and Homestead. These last two are still somewhat popular but less productive than modern hybrids like Champion. Marglobe continues to produce large, thick-walled fruits with excellent flavor. Marglobe is listed as a determinate but will easily fill a 6-foot cage.

Moskvich (Moskovich) (3, 14)—60 days—is an early-maturing Russian heirloom producing 4- to 6-ounce fruits with luscious flavor. This is one you can plant early since it is tolerant of cold, wet spring weather.

Nyagous (1, 11, 19, 20)—80 days—is a blackish red tomato that looks like an oversized plum. For a nonhybrid variety it produces large clusters of sweet fruit with lots of complex flavor. This one is destined to be a market favorite if people will just try it. Plants are large and will need serious staking.

Oregon Spring (3, 6, 8, 12, 13, 17, 19, 20)—58 days—is a cold-tolerant variety developed for the Northwest, but it has done well in our trials in

Marglobe was a favorite in Grandma's garden. Chances are you will like it, too.

Moskvich, an early-maturing Russian heirloom, is worth a try in the Texas tomato patch.

Nyagous is a Russian black tomato with a sweet, complex flavor.

Surprisingly, this northwestern variety, Oregon Spring, did well in Houston variety trials and scored well in taste tests, too.

Harris County. The plant is determinate, and the fruits are juicy and delicious with comparatively few seeds. Cold tolerance and heat tolerance seem to be linked, so this one is worth trying, especially early in the season.

Persimmon (1, 10, 13, 15, 17, 19, 20)—80 days—is one big heirloom orange-yellow tomato that's worth a spot or two in the garden. It is not a mild, sub-acid type. This one will lap over a burger and has excellent flavor. Total pounds per plant may not be too impressive, but these 1- to 2-pound beauties are worth it.

Purple Calabash (17, 19)—80 to 90 days—is a large fruit with dark, purple-black skin color and ruffled shoulders. The flesh is redder with a raw meat look but very sweet and loaded with complex tomato taste. Like a number of heirlooms, you will likely discard some of the ugly to

Total production might not keep up with the hybrids but Persimmon is worth growing for its delicious and large fruit.

get to the good stuff; but this one's worth a try for how good it is on a sandwich. Purple calabash might be a good candidate for grafting to a disease-resistant rootstock to boost production.

Rutgers VFA (1, 8, 14, 17, 19, 20, 22)—75 days—was a regular in Grandma's garden. Even though it was developed in the 1920s, it's still worth a try if you like great-tasting heirlooms. Somewhat determinate, it may grow out of your cages by season's end, but you'll love these 6- to 8-ounce tomatoes that are wonderfully fresh and superb for canning.

Sioux (1, 13, 19)—70 days—is an old standby for Southern gardens. Chances are it was one of your grandmother's favorites. It grows well in hot weather and has thick walls with acidity and flavor. Super Sioux is an improved version.

Purple Calabash is a dark purplish red heirloom with complex tomato flavor and a juicy, melting-textured flesh.

For trial: Black, Black Russian (10, 17, 18, 19, 20), Black from Tula (1, 11, 14, 19, 20), Black Prince (1, 3, 6, 15, 19, 20), Japanese Trifele Black (11, 17, 19, 20), Paul Robeson (1, 13, 15, 19).

CHERRIES AND SMALL-FRUITED VARIETIES

Small-fruited tomatoes are tremendously productive and continue to produce long into the summer when the large-fruited varieties have given up or succumbed to spider mites. You may still have to spray for insects, diseases, and mites, but the flowers will set in spite of the heat. The really tasty varieties like Sungold or Sweet 100 could even be planted as a trap crop for leaf-footed bugs—they love them. Actually, all of the cherries are so delectable that most any will tempt the leaf-footed bugs and stinkbugs. You can concentrate your sprays on these plants and spray less on the larger-fruited varieties. Since these potential trap crops are still edible food crops, you have to use pesticides that are labeled for use on tomatoes and observe the waiting period before harvesting.

One or two plants per family are enough to keep you harvesting until you're ready to give up. If you have too many plants, they tend to be neglected—attracting pests and dropping seed to sprout up as volunteers next year. The tiny-fruited variety Texas Wild is scrumptious, but it is also notorious for becoming the new weed in the garden.

This group includes the best candidates for container gardening. Smaller fruits are less susceptible to blossom end rot caused by a water shortage at the blossom end of the developing fruit. However, most are fairly susceptible to cracking. Cracking occurs when there is a slowdown in growth—usually because of dry weather or heat stress—although low fertility could also be to blame. After a rain or thorough watering, the fruit expands faster than the skin, and the skin cracks. Side-dressing with fertilizer or foliar feeding might also be a contributing factor if the plants have been stressed for nutrients. Regardless, don't skip on fertilizing—just make sure to be regular about it. All are indeterminate and rather vigorous unless otherwise noted.

Small cherry tomatoes are popular at Boggy Creek Farm in Austin, Texas.

Enchantment Hybrid VFFN (10)—70 days—is a 4-ounce plum-shaped tomato on a determinate plant. Great for cooking or canning, it also has good flavor for fresh salsa or a salad.

Gardener's Delight (8, 19, 20)—72 days—produces long clusters of bright red, delicious fruits. This is a large indeterminate plant that demands a lot of picking.

Enchantment is another Roma look-a-like with flavor, disease resistance, and good productivity.

Golden Rave Hybrid FT (2, 3, 7, 16, 20)—67 days—is a sweet elongated tomato in the 2-ounce size. Plants are strong growing and productive with resistance to cracking.

Health Kick Hybrid VFFASt (6, 7, 19, 20)—72 days—is a saladette-type (elongated plum type, meaty with less juice) tomato with excellent flavor and lots of health-promoting lycopene. Plants are determinate with 3- to 4-ounce fruits. This would be a good candidate for a patio planter.

Jaune Flammee (11, 20)—80 days—is larger than a cherry but far from a slicer (about 3 ounces or apricot sized). This French heirloom is a yellow to orange tomato with flavor. In fact, it really zings unlike most yellow tomatoes that are touted as low acid (which means bland flavor!). This is great for salads, or if you summer at a French villa, you may find them on your breakfast plate—or so a delighted farmers' market customer mentioned when she discovered them at my table. She came back every week, and, yes, this is a hint, if you're considering heirlooms for market sales—include this one. It's a very productive plant that is a bit susceptible to early blight, but most heirlooms are.

Jolly (19, 20)—75 days—is a Ping-Pong ball–sized fruit often with a bit of a nipple. It's a 2001 AAS winner, quite productive, and very sweet. It does crack some in our fickle Texas weather.

Golden Rave Hybrid tomatoes are 2 ounces of tasty yellow goodness on a very productive and vigorous plant.

Health Kick is a saladette-style tomato—full of flavor and beneficial lycopene.

Jaune Flammee is a French heirloom with great flavor and excellent productivity. It might look like a mild-mannered, orange-skinned tomato, but it has a zingy, complex flavor.

Juliet Hybrid (2, 3, 10, 19, 20, 21)—60 days—a 1999 AAS winner, is one of the few tomatoes that are successful as a fall tomato, either as a plant babied through the summer or one set out specifically for fall in June or July. Small-fruited varieties tend to set earlier in the heat and ripen more rapidly in fall with the shorter days and cooler weather. Juliet has an elongated fruit—looking somewhat like a skinny Italian heirloom but with clusters of sweet fruit on an indeterminate plant.

Margherita Hybrid VF (7, 19, 20)—70 days—resembles an elongated Roma, but it has much more flavor and firm texture that isn't grainy, as Romas can be. It is a relatively dry tomato that is perfect for salsas, and the seeds are easy to remove, if so desired. It is also great for broiling—and super on a pizza! Fruits are produced on a determinate plant that is easy to cage.

Porter (1, 19, 20)—70 days—was a regular addition to Grandmother's garden fifty years ago and is still available. You could call it the original grape tomato. It was a bit large for a grape, about 2 ounces, and it isn't real punchy in the flavor department, but Porters would set all summer in the heat and tasted pretty good when every other variety had stopped producing. The Porter Seed Company in Stephenville was a mainstay for Texas gardeners for many years.

Juliet is another small saladette-type tomato that has been a good producer in the fall garden.

Margherita is a great, productive tomato for fresh salsa. It has flavor but not a lot of juice and seeds, so it is excellent if you don't want wet salsa.

The Porter tomato isn't a real zinger for flavor, but it produces in the heat as well as in the fall.

Principe Borghese (1, 15, 17, 19, 20)—75 days—
is a plum-shaped tomato prized by the Italians
for drying. It is rather dry when fresh and is
produced abundantly on a determinate plant.

Snow White (19)—75 days—was a real surprise
in the garden. White tomatoes have not been
too impressive in most trials—the big ones are
ugly and often lack flavor. This little indeterminate white cherry (best when pale yellow) is
sweet and has enough acid to make it interest-
ing. To keep it from being ignored at your farmers' market table, try packing it in snack-size
plastic bags with other cherries like Sungold or
Sweet 100. This makes for a fun treat that real
veggie lovers can't resist.

Sungold Hybrid (widely available)—57 days—
may be the most delicious tomato yet. It's
sweet, it's tart, and it has complex tomato flavors—if it didn't roll off a sandwich, it would
be a mainstay in the Adams garden. It also has
a tendency to crack after a good rainstorm or
any event—like side-dressing with fertilizer or
foliar feeding that spurs rapid growth. There
is always room for one plant though, and we
can only hope someday the plant breeders will
get these genes into a slicer. The seed producers have certainly not ignored it. Varieties like
SunSugar Hybrid and Super Suncherry Hybrid
are worth trying.

*Valued for drying, these
tasty Principe Borghese
Italians are a fun addition
to the tomato patch.*

Snow White cherry tomatoes were a real surprise. Sweet with a touch of acidity, they are an extremely productive novelty.

Super Sweet 100 Hybrid VF (widely available)—65 days—is tomato candy. Huge clusters of small 1-ounce fruits and a very large plant are its claims to fame. Plant this one for the little children, but don't be surprised if you crowd them out of the way to get your share.

Sweet Chelsea Hybrid TFN (8, 16, 19, 20, 21)—67 days—produces a large cherry with great flavor and productivity. This plant is a strong grower with disease resistance. In our trials in the Houston area, we produced over 60 pounds per plant—that's a lot of "pickin'."

Texas Wild (similar to Mexico Midget, Matt's Wild, and Wild Cherry) (1, 3, 11, 13, 19, 20)—65 days—is one of the tomatoes that to most gardeners is seemingly too small to be worth the effort. The flavor is outstanding though, and kids love them. They could also become a bit of a weed in your garden, as they tend to grow over things, and after you give up picking in frustration, they reseed everywhere. These tiny tomatoes would be good candidates for a 20- to 30-gallon container on a patio or in the garden maybe; start with one plant.

Viva Italia Hybrid VFFNASt (17, 19, 20)—72 days—is touted as a hybrid paste tomato but is really tasty enough to slice up for a sandwich if it weren't a bit on the small side. The texture is firm, not grainy like most Roma

Sungold may be the best-tasting tomato in the world.
Too bad it's too small for a BLT, and it has a tendency to crack.

types, and is also larger than a Roma. Viva Italia is likely to be a mainstay variety in your garden if you like fresh salsa and a superb canning variety. It is susceptible to early blight but often recovers for a late summer/fall crop.

Cherry tomatoes like Sweet 100 and Sungold are a favorite at farmers' markets.

Sweet Chelsea is one of the most delicious and productive of the cherry tomato varieties.

(opposite) The Texas Wild tomato literally bursts with intense tomato flavor—it's almost tomato candy!

Viva Italia looks like a Roma but has more flavor and is more productive and disease resistant.

For trial: Babycakes (2), Isis Candy (1, 11, 17, 19, 20), San Marzano (3, 12, 13, 16, 17, 19, 20), Suncherry Hybrid (3, 19, 20), SunSugar Hybrid FT (8, 19, 20, 21), Sweet Baby Girl (6, 7), Tami G Hybrid (grape type) (16).

Insects, Diseases, Weeds, and Varmints

Tomato Pests

Gardeners are not the only ones who love tomatoes. Insects and critters love the tomato's juicy sweetness, diseases thrive on the nutrient-rich plants, and weeds flourish in the high fertility of the soil where tomatoes grow. In a good year, when spring rains don't fall too frequently and you plant early to avoid the leaf-footed bugs, you will likely harvest a good crop without any spraying. Frequent rains favor the early blight fungus, and by late July the bugs are thriving. A rainy spring almost ensures that you will need to apply a preventive fungicide for disease control. Once the disease is flourishing with lower leaves turning yellow and then brown, it's too late to control the disease in time to salvage much of a crop. There are some low-toxicity and organic fungicides that can help to save the day. Neem oil, sulfur, and a new bacillus—*Bacillus subtilis*—should get the job done in all but the worst years.

New gardeners and especially those determined to grow organically often suppose that growing organically means no spraying. In most cases, it just means spraying more often with less toxic materials. Of course, there are other factors. If the gardener is serious about enriching the soil with compost and carefully monitoring the garden to keep ahead of pest problems, he or she may, in fact, grow a super crop of tomatoes with little or no spraying. Regardless, it pays to know what sprays are available—chemical and organic. Just remember that it's naïve to presume that just because you use compost, the bugs and diseases will stay away.

If you follow the label directions, including the days to harvest (after spraying), you don't need to worry about the quality of tomatoes you harvest. Even with so-called chemical pesticides, the levels are so low at harvest time—assuming you use the correct rate of application and wait the required days to harvest—as to be inconsequential. Remember that it's the dose that counts. Many of the medicines we take are toxic if taken at high rates.

Knowing where to turn to deal with pests and diseases is another important factor. Master Gardening is a volunteer program of Texas A&M University's AgriLife Extension. Master Gardeners who graduate from the program are available at many local AgriLife Extension offices to answer most of your gardening questions via the telephone, or you can usually bring samples in for analysis. These services are more extensive in the large metropolitan centers, but Master Gardeners are spread across the state. County extension agents can assist with the more difficult or commercial questions, and the Texas A&M specialty staff (entomology, pathology, and weed science) are available via the Internet. Check with your local county extension agent about how to submit samples (good digital photos are often adequate). The following Web sites will prove invaluable:

http://aggie-horticulture.tamu.edu/,
http://insects.tamu.edu/extension/,
http://plantpathology.tamu.edu/extensionprograms/index.htm,
and http://soilcrop.tamu.edu/research/weeds/index.html.

Most gardeners take a commonsense approach and use a range of pesticides and fertilizers, both organic and inorganic, with emphasis on low-toxicity materials.

You need a plan if you want to grow a pest-free tomato crop with a minimum number of sprays. The following list summarizes the steps in planning your garden.

1. Prepare a fertile soil bed, rich in organic matter, with a good fertility base. It's best to fertilize based on a soil test, but 1 to 2 pounds of a complete fertilizer like 15-5-10 per 100 square feet is a good starting point if there is not time for a soil test.

2. Soil solarization can significantly reduce pests in the next season's crop. It needs to be done in the late summer (July–September) when the sun is blazing and the temperatures are intense. Work up the soil, moisten it, and then cover with ultraviolet (UV)-resistant clear plastic—this is the type used for greenhouse covers and usually is not available at local hardware stores. If possible, befriend a greenhouse grower and see if you can buy some scraps. An alternative may be to use a double layer of regular clear plastic since you need six to eight weeks of coverage to eliminate a number of pests. A single layer of non-UV-resistant plastic will break down too soon.

3. Ensure good drainage with raised beds, or work the soil into ridges 12 to 18 inches high and plant on the ridges.

4. At planting time, wrap the tomato stems with a cardboard collar—½ inch beneath the soil and 1 or 2 inches above. Use staples to secure it. A collar of aluminum foil can be used and is a bit easier to work with. Even a 2-inch section of cardboard from the center of a paper towel roll will work if you cut it up one side to facilitate placing it around the stem.

5. Use a starter solution at planting. Any soluble fertilizer used about one-third to one-half strength with a pint per plant should suffice. You can also use slow-release fertilizer tablets in the hole. If you like to experiment, inoculate the root zone with mycorrhizae or similar beneficial microbes.

To solarize a garden bed, first work up the soil. Then water it until it is moist, and cover the area with UV-resistant plastic for six to eight weeks in the heat of summer—usually August and September. Many of the weeds, insects, and disease organisms will be reduced.

6. Use a good, sturdy cage—reinforcing wire is a minimum—and be sure to add two stakes, such as 2 × 2s or T-posts, to keep the cage from falling over in a windstorm. Wrap the cage with fiber row cover, and secure the cover with clothespins—this can keep the insects out or keep them safe and sound if they are already there (not good). Just be sure to check for pests before wrapping with row cover. When the plants are knee high, remove the cover, wash (if necessary), and store it for next year.

7. Be prepared for unexpected spring and fall frosts. Spring frosts were covered more thoroughly in the planting transplants section, but to review, cover plants with row cover, paper caps, or an old milk jug (bottom removed) to get past those untimely frosts. An early fall frost can be equally disappointing. It can shut your tomato production down before the fall crop has sufficient time to begin ripening. Then you are faced with a bumper crop of green tomatoes and a lot of green tomato chowchow.

8. Consider a preventive spray program to stop fungal diseases before they ravage the plant. Be sure to spray the undersides of the leaves as well as the tops. Check with the AgriLife Extension office or Aggie Horticulture (http://aggie-horticulture.tamu.edu) for spray recommendations. Remember that you can remain organic by using neem oil, wettable sulfur, or one of the new biological fungicides to prevent diseases like early blight. There are also a few hybrid varieties with resistance to early blight (look for an A indicating resistance to *Alternaria solani* on the alphabet list following the variety name). Mulching with a combination of newspapers covered with alfalfa hay will also reduce the splashing of fungal spores from the soil.

9. Examine your plants carefully a minimum of several times per week to spot early infestations of hornworms, aphids, spider mites, stinkbugs, and leaf-footed bugs. Handpick and destroy overwintering insects like the large leaf-footed bugs to reduce their early buildup. Then spray early and only when necessary with the least toxic pesticide that will do the job. Sometimes you have to spray more often to get the same control. Consider planting a trap crop like cherry tomatoes or southern peas in a far corner of the garden. Bugs will concentrate here, and you can spray more often (still using pesticides labeled for the trap crop).

10. Tomatoes don't have to ripen to full color to have good flavor. Once they have reached the half-color stage, you can harvest and let them finish ripening inside away from marauding pests.

INSECT PESTS

Aphids or plant lice are a common pest in the early spring when temperatures are cool. These tiny pear-shaped insects usually cluster on the tender growing tips of the plant and also like to hide under the leaves. They may be red, green, or yellow in color, and the introduced fire ant may be protecting them—actually farming them—for the honeydew they excrete. Aphids damage the plant by sucking out the juices, and some may be disease vectors (carriers).

Sprays are often not effective because aphids develop resistance too quickly. Most are females, and they are born pregnant. If a few are resistant and survive, before you know it, you have a large pesticide-resistant population. High-pressure water sprays will often get most of them; lady beetles can finish off the rest.

These small pear-shaped insects come in a variety of colors. Aphids usually hide on the undersides of the leaves and feed by sucking the juices out of the plant.

Beetles can damage tomatoes on occasion. Usually they are a problem on small plants where a number of tiny holes in the leaves are an indication of their damage. Dusting with the garden pest-control form of diatomaceous earth (microscopic shells of ancient sea creatures whose tiny, sharp points injure the insects) or a registered insecticide will usually get rid of enough beetles to get the plants established. Once the plants begin making vigorous growth, any damage is relatively insignificant.

Cutworms can bring disappointment early in the season since they take down your transplants within days of planting. These caterpillars crawl along the soil surface and munch on stems at ground level. They are most active at night, so you may not even know what to blame. During the day they may be rolled up under a dirt clod and hard to find. Chemical controls are not very effective since the cutworms don't eat much and do their damage before the pesticide can take

A heavy waxed paper or cardboard wrap around the tomato stem at transplanting will discourage cutworms.

Grasshoppers are more of a pest in rural areas, but they can do a lot of damage anywhere when they are hungry and numerous. They often feed on ripening tomatoes.

effect. A physical barrier is the easiest way to thwart them. Wrap a cardboard or aluminum foil barrier around the stem extending ½ inch into the soil and as much as 2 inches above the soil line to keep them from getting at the stem.

Grasshoppers are not much of a problem early in the spring, but later in the season when they get large and develop a taste for juicy tomato fruits, they become a real pest. A full-size grasshopper is virtually immune to pest controls that would be available to the home gardener. In rural areas, where they tend to drift in from the surrounding pastures and farmland, they become a never-ending challenge by midsummer. Spreading Nosema bait early in the season while the grasshoppers are still small and susceptible to the disease may help. There are also baits with rolled oats treated with an insecticide. This is fairly host specific since the grasshoppers eat the bait and beneficial insects don't.

Leaf miners are the larvae of a small fly tunneling between the cell layers of the tomato leaf. The damage looks like squiggly white lines in the leaf. It's hard to get to the leaf miners except with a foliar systemic, and the home gardener does not have access to these types of pesticides for use in the vegetable garden. Persistent spraying with a contact pesticide to kill the adult fly before it lays its eggs is one solution, but in most cases the damage is more disconcerting than destructive.

Spider mites love tomatoes and are especially prevalent in dry years. Rainfall and even sprinkler irrigation may tend to reduce their numbers; unfortunately, rainfall and irrigation increase the incidence of disease. Spider mites are tiny, so you need a hand lens to observe them. Another technique to use if you are new to mite hunting is tapping the leaves on a piece of white paper—then check to see if the dust specks crawl off. When populations really become elevated, the plant leaves may be covered with fine webbing and the foliage will have brown, stippled areas where the cells have died. One of the best solutions is to use a water wand with high pressure to knock them off and give natural predators a chance to clean them up. Do this early in the morning, and the foliage will be dry by evening when cooler temperatures make disease infection more likely. Wettable sulfur and neem oil will reduce spider mite numbers, but use them separately unless you have tried the combination on a test plant, since together they may be more likely to burn the foliage—especially in hot weather when mites tend to be a problem. Insecticidal soaps can also be effective, but again there is the potential for leaf damage (phytotoxicity) during the heat of the day. This damage is most likely to show up as brown areas on leaf edges and tips where the spray concentrates. In general, most pesticides are best applied early in the morning or late in the afternoon. Once plants have reached the point where the foliage is covered with webbing, it is usually best to pull them out and plan for another day.

When mite populations get so high that you begin to see webbing and the foliage becomes stippled as the cells die, the mites have won and it's time to pull up and discard the plants.

Stinkbugs and leaf-footed bugs don't wipe out the tomato crop, but they can sure mess it up. These pests are true bugs with mouthparts used to suck the juices out of the plant. They tend to prefer fruits and seed pods, so they do the most damage to the plant parts we like to eat. Stinkbugs are shield shaped and vary in color from green to brown. There are beneficial predator stinkbugs that attack other insects, but they are rarely effective in the home garden. The leaf-footed bug is one of the rogues in the tomato patch. This insect is elongated with leaflike appendages on the back legs. It's brown to dark brown in color with a white stripe across the back. They can often be found clustered on the ripening tomato fruit. By the time you harvest at the vine-ripe stage, there will be white corky areas under the skin that reduce the palatability of the fruit. Spraying with neem oil/pyrethrin sprays can be effective, but you may have to keep at it weekly to get good control. Part of the problem is that this pest has a wide host range. You can kill them in the garden, but new ones may fly in every day from your pecan tree or from weed hosts. New gardeners with a bit of knowledge about the insect world often think they have hit the mother lode of beneficial insects when they find a bunch of leaf-footed bug nymphs on the plants. The nymphs are orange and look like they could be baby assassin bugs. You never see a lot of predators like the assassin bug clustered together—they're a tad cannibalistic—so you're seeing baby leaf-footed bugs. Spray them with a registered pesticide or dust them with a diatomaceous earth labeled for garden use.

Worms and more worms (really caterpillars—earthworms in the ground are a sign of good soil health) seem to find their way into the tomato patch. One of the most sinister looking—if you can see it—is the

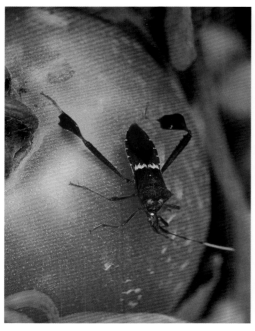

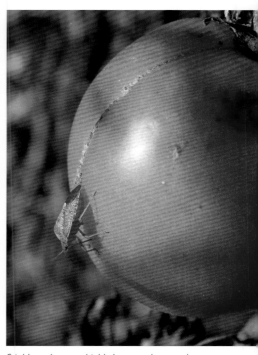

The leaf-footed bug is one of the most damaging and persistent pests on tomatoes. It sucks juices out of the plant—mostly the fruit—causing cells to die and a white corky layer to develop.

Stinkbugs have a shield shape and cause the same damage as the notorious leaf-footed bug.

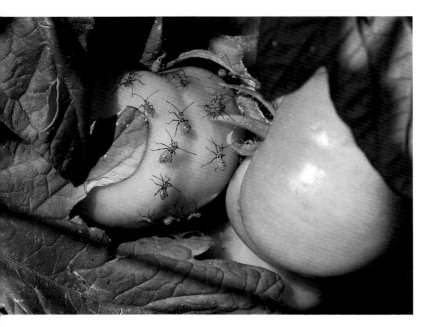

Masses of small orange bugs may have you jumping for joy in hopes you have a family of assassin bugs (a predator) in your tomatoes. When these tiny bugs are numerous, they are almost always plant feeders—in this case, young leaf-footed bugs.

tomato hornworm. These caterpillars are big but also the color of tomato foliage and can eat a lot of foliage before their bulging finger-sized shape gives them away. Many times the first symptom we notice are sections of the plant where only a few tough stems remain. The horn on this big caterpillar is a bluff—it can't sting. One control solution would be to spray or dust the plants with *Bacillus thuringiensis* (Bt) to take the baby caterpillars out while they are actively feeding. Susceptible caterpillars eat the Bt, get sick and stop feeding almost immediately, and then die in a week or less. It's too late to control the big ones this way. They will have to be controlled mechanically, such as by handpicking. The tomato fruit worm (also called the cotton boll worm or corn ear worm) is another potential pest. It's much smaller than the hornworm but every bit as ravenous.

Imported fire ants are a common pest in the garden. Even though considered a predator, they do directly attack some plants like okra. While

Tomato hornworms are easy to control when small, but they are so well camouflaged they often eat a lot of foliage before they are discovered.

Tomato fruit worms have a wide host range. They are also referred to as the cotton bollworm and the corn earworm.

they may not munch on your tomatoes, they do like an occasional ankle, and they also farm aphids that may be attacking your tomato plants. Ant baits with spinosad are safe to use in the garden, and you can use other baits and insecticides outside the garden.

A useful tool in your arsenal for monitoring pests is the sticky trap, which is frequently used in the greenhouse to identify insect pest outbreaks. The yellow color attracts the critters, and a sticky substance immobilizes them so the grower can determine when the population is expanding rapidly enough for control measures to be necessary. They can be used outdoors as well. They don't catch enough insects to amount to a control technique. They just let you know when spraying is necessary.

The imported fire ant is primarily a predator; unfortunately, it also "farms" aphids—protecting them and feeding on the honeydew they excrete.

A yellow sticky trap in a greenhouse tomato-growing operation is used to capture plant pests. The trap does not affect control, but it signals the grower when it is necessary to spray.

BENEFICIAL INSECTS

What garden is complete without beneficial insects like lady beetles? To a large extent it's a measure of the health of your garden and your personal stewardship if you can keep these helpers involved while checking the buildup of plant pests. Releasing predators and parasites is not always very effective. They have a tendency to move on to the neighbor's yard. They can be quite effective in a greenhouse where they are somewhat confined. Also, releasing them at night may cause them to settle down some and not move on as rapidly. Of course, if you don't have a population of their favorite foods like aphids, white flies, mealybugs, or caterpillars, they may decide to shop elsewhere anyway. In addition to lady beetles you may want to try green lacewings and predatory wasps. Actually wasps are usually rather abundant in Texas gardens. We don't

Releasing beneficial insects frequently fails because they often move elsewhere in search of food or a more suitable environment. Lady beetles in the greenhouse— because they are confined— often do a good job of cleaning up aphids and other pests.

This hornworm has been parasitized by a tiny wasp.

The wheelbug is a beneficial predator insect.

want them nesting over the front door, but there's certainly no need to kill every wasp that flies—they are all predators or parasites. Other "good" bugs include assassin bugs, wheelbugs, ground beetles, syrphid flies, praying mantids, big-eyed bugs, soldier beetles, spiders, and beneficial nematodes (these last two are not technically bugs).

Although you can purchase beneficial insects (sometimes available at garden centers, but usually ordered online or via catalog), as previously mentioned, they often go astray. Perhaps the best way to get help from these insects is to plant flowers that attract them with nectar and pollen. Some of the best include cilantro, dill, yarrow, black-eyed Susan, alyssum, and dwarf sunflowers.

TRAP CROPS

As previously mentioned in the section on cherry tomatoes, leaf-footed bugs and stinkbugs are really drawn to these little sweeties. You do not want to spray them with anything that is not labeled for tomatoes, but at least you can spray often and use products you may be leery of on the rest of the crop. Be sure to look for beneficial insects like lady beetle larvae or green lacewings. When they are present in good numbers, a few low-toxicity sprays with neem oil or insecticidal soap may be all that's necessary to control the pests in the trap crop. Always plant trap crops away from the main crop so you don't attract more pests than you lure away. Can you eat the trap crop? Yes, you can, if you have sprayed materials registered for use on the tomatoes. If cherry tomatoes like Sungold aren't bait enough, then try planting some southern peas in the same area. Aphids and stinkbugs love these, too. The large sunflowers that are grown for their seeds are another magnet for leaf-footed bugs and stinkbugs.

Small-flowered marigolds and cereal rye (like the variety Elbon) are to some extent trap crops for nematodes. The nematodes cannot feed effectively on them, and the green manure you turn under (especially with the rye) will also reduce their numbers.

Squash plants are more attractive to white flies that transmit the tomato yellow leaf curl virus and could be used as a trap crop for this vector pest should this disease become a serious problem in your garden.

Ornamental millet like Jester Hybrid (7) makes a beautiful trap crop for stinkbugs and their relatives. You can spray with pesticides approved for ornamentals, and the birds will likely benefit from the seeds later in the season when the pesticides have broken down. Of course, you can also use pesticides registered for tomatoes and organic products to get rid of the six-legged demons. Dill and lovage (a celery relative) often prove more attractive to the tomato hornworm than the tomato does. As a bonus, the caterpillars are probably easier to spot in open dill foliage.

Chile pequins and other small-fruited peppers are great for pepper sauce, but they also seem to satisfy the mockingbirds that otherwise would be pecking holes in your luscious tomatoes. Radishes can be used to lure away the aphids early in the season, and you still get to use the radishes in a salad. The best control is a high-pressure water spray followed with an insecticidal soap spray, so there are no toxins to worry about.

TOMATO DISEASES AND ENVIRONMENTAL STRESS

Early blight (*Alternaria solani*) has disappointed many a first-time tomato grower. You think everything is going great—the plants are vigorous, it rains twice a week, life is good, and daily you're adding to your file of tomato recipes. Then the first lower leaves start to turn yellow. No problem; "look how many green leaves I still have," you note. But a week later more leaves are yellow, and the first yellow ones now have turned brown. Early blight has the upper hand; if you don't move quickly to stop its spread, you'll be buying your fresh tomatoes at the farmers' market. This fungal disease has been spread through the splashing of spores from the soil. It got its start on lower leaves, and now the fungus is producing clouds of spores with every rain or sprinkler irrigation. Using mulch to cut down on the splashing of spores would have helped. Removing the lower leaves 12 to 18 inches up from the soil (by the time the plants have grown waist high) would have improved air circulation and allowed the leaves to dry more quickly—this makes them less susceptible to infection. Spacing the plants 4 feet apart instead of 3 feet would also have helped with air movement. And finally, applying a fungicide to the plants early in the season—making sure to spray the undersides of the leaves as well as the tops—would have reduced the degree of infection. If early blight is caught early, when a third or less of the foliage is showing yellow leaf symptoms, you may still be able to arrest the disease enough to save the crop with fungicide sprays. When a half or more of the foliage has turned brown and the rest of the foliage is yellow, it's time to pull the plants and try for a fall crop of cherry tomatoes. Cultural techniques may help, but with persistent diseases like early blight, even if you concentrate on varieties with some resistance, in a rainy year you just anticipate problems with this disease and spray to reduce its prevalence.

Blossom drop is common when temperatures are consistently above 75 degrees Fahrenheit at night and in the mid-90s during the day. Cherry tomatoes and small-fruited varieties are less susceptible to blossom drop, and there are a number of "hot-weather" varieties offered in catalogs. Typically the large tomatoes, even if touted as hot-weather tolerant, produce tomatoes that are smaller than tomatoes produced on the same variety early in the season.

Septoria leaf spot (*Septoria lycopersici*) is another fairly common fungal disease on the leaves and stems of tomatoes in a wet year. These spots are small and dark brown with tan centers and also start on the lower leaves. Adding insult to injury, you

Early blight is one of the most common fungal diseases on tomatoes. It starts on the lower leaves and rapidly causes the leaves to yellow and turn brown.

Tomato blossoms often drop when temperatures are in the 90s during the day and above 75 degrees at night.

could have both early blight and septoria. The early application of preventive fungicides is the key to stopping this disease before it gets started. Removing old plant debris and cultural techniques like wider spacing and good air circulation also help. Rotate tomatoes to another section of the garden at least every four years.

Bacterial leaf spot (*Xanthomonas campestris* pv. *vesicatoria*) is most common in wet years or in plantings with overhead sprinkler irrigation. It is also a pest of peppers and is often more severe on this crop. Avoid the use of sprinklers, and spray with a copper fungicide during and after wet seasons in the garden.

Blossom end rot is not caused by an organism; it is an environmental response. It can happen as a result of dry weather, occurring in drought-stressed plants in most any garden or in raised beds, but excessively wet weather can also cause it to occur. The blossom end of the fruit becomes dark green and water soaked first; then it may become black and leathery—eventually fruit-rotting organisms may be involved. This condition is most likely to occur with tomatoes in a container, especially medium- to large-fruited varieties, or it may show up in a planting with heavy, poorly drained soil. The developing cells at the blossom end of the fruit die as a result of water- and calcium-deficiency stress. This can happen during a dry spell or if you allow the plant to wilt severely in a container. It can also happen after a drenching rain when the soil becomes waterlogged; the roots cannot take up enough water because the soil atmosphere is reduced (the open spaces that normally contain air are filled with water), so the roots are deficient in oxygen, leaving them unable to take up the water that is all around them. Good calcium nutrition seems to help the plant deal with the situation, so plants are sometimes fertilized or sprayed with soluble calcium materials like calcium nitrate or calcium chloride. The best cure is even watering. Use low-volume irrigation in the garden, and set plants on ridges for good drainage. Mulch can help conserve moisture during dry spells

Container-grown tomatoes are especially susceptible to blossom end rot due to the potential for water stress.

and keeps the moisture availability more even. If you grow tomatoes in containers, use at least a 20-gallon container and install a drip system to keep them watered while you are at work. Note: A container-grown tomato with a good set of large fruit may need water two to three times per day during the early summer as tomatoes are beginning to increase in size.

Tomato fruit rots can be caused by several fungal and bacterial organisms. Phoma rot (*Phoma destructiva*), *Phytophthora,* and alternaria fruit rot are the most common fungal organisms, though bacterial fruit rot (*Xanthomonas campestris*) can be a concern in extremely wet years. Avoiding the use of overhead sprinkler irrigation is one key to reducing the potential for fruit rots. In wet years preventive fungicides may be necessary to keep these diseases in check, and copper fungicides will have the added benefit of reducing bacterial infections.

Both fungal and bacterial organisms can cause fruit rot. It is especially prevalent when the fruit has soil contact and when the growing season is wet.

Waterlogged soils can cause wilting because of lack of oxygen in the soil and the resulting inability of the roots to take up water.

Nematodes are microscopic round worms that attack the root systems of plants for food and nutrients. In the process they may cause the roots to form nodules. Other plant pathogenic nematodes may just damage the roots and cause them to be stubby and less efficient or fewer in numbers. In any case, the end result is a plant that is suffering from a lack of water and nutrients. The plants may be stunted or yellow when small, or if the nematodes build up later in the season, the plant may have a decent crop but wilts prematurely and eventually dies before maturing the crop to

Root-knot nematode damage on tomato roots.

full size and quality. There are no postplanting chemical nematicides, so the best you can do is to plant nematode-resistant varieties or graft susceptible varieties on resistant rootstock. Making yearly applications of compost will help encourage microorganisms that can reduce nematode populations. One advantage of gardening in a heavy clay soil is that nematodes usually are not so much a problem as in sandy soils.

Viral diseases are systemic and cannot be controlled once they have infected the plant. Some are carried by insect vectors, and some can be transmitted by mechanical damage—just working with the plants after handling an infected plant can spread the virus. Any plant that appears spindly with shoestringlike growth or a mosaic pattern to the leaves is best rogued out. Herbicides can cause similar symptoms, but the damage is more general, while viral infections tend to show up initially in a single plant or just a few scattered plants in the garden.

When a uniform group of plants shows symptoms typical of a virus, they may be suffering herbicide damage—in this case caused by hay mulch contaminated with herbicide

Tomatoes with twisted growth, shoestring-like tips, and a mosaic pattern in the leaf are likely infected with a virus and should be rogued out of the garden.

Sunscald and cracking are common in the Texas tomato patch. Look for varieties that claim "good foliage cover" if sunscald has been a problem, or plan to cover the planting in July with fiber row cover or shade cloth. Cracks can be a result of several factors. A slowdown in growth followed by a rainstorm or side-dressing with fertilizer can cause a spurt of

Tomato varieties with limited foliage cover can leave the fruits susceptible to sunscald.

Some varieties like Sungold are notorious for cracking after spurts of growth caused by rainstorms or added fertilizer.

These cracks at the stem end of the fruit are due to stress from heat and moisture during the mid- to late summer growing season.

growth that causes rapid expansion of the fruit's interior; the skin cracks because it cannot keep up with the interior expansion. Some varieties, like Sungold, are notorious for cracking, but under the right circumstances almost any tomato can crack. Stress cracks are also common late in the season as extreme heat becomes a factor.

ORGANIC AND LOW-TOXICITY SPRAYS THAT REALLY WORK

In deference to the rapidly changing state of pesticide registration, you are on your own for chemical pesticide recommendations. Fortunately, most of this information is available on the Internet, or you can contact your local AgriLife Extension office. You can also review the pesticide labels at a local nursery. What you are likely to find is that there are not many so-called chemical pesticides available for your purchase. This may leave us somewhat at a disadvantage compared to our gardening parents and grandparents, but let's face it, there were some careless pesticide abuses in the good old days.

Responding to calls from gardeners for thirty-plus years made me realize that if a pesticide could be misused, it was—usually by the other spouse in the house. The excuse for misuse went something like this: "Somebody put chlordane on my tomato bed. How long do I have to wait before I can start eating the fruit again?" In the 1960s and 1970s chlordane actually could be used in the vegetable garden with restrictions, but toward the end of my career as a county extension agent for horticulture it wasn't even allowed for termite protection. What usually happened was someone pulled out an old bottle of chlordane from the garage and treated around the house for termites. It just so happened that the tomato bed was located next to the house as well. If the chemical wasn't labeled for use on tomatoes, then there was no USDA-recommended "waiting period prior to harvest." The difficult answer was "destroy the plants and plant new ones somewhere else."

So, is the organic alternative viable for home gardeners? Of course it is. You may lose a few battles, but you will be safer using low-toxicity pesticides, or you may get by with not spraying at all. You often hear the following adage from organic gardeners: "I just plant extra and let the bugs take their half." In my experience, the bugs sample half of everything. Maybe trap crops tend to encourage the critters to confine their feeding to "the extra half," but in a bad pest year we often have to intervene with some type of spray program. Probably the real "pesticide danger" is of more concern to the applicator—the person spraying, drenching, or dusting—than to the consumer of the produce. Avoiding spills, clothing contamination, spray drift, and a host of other unplanned mishaps is far more likely to be a problem than consuming the tomatoes after waiting the required period before harvest with chemical pesticides that are labeled for use on tomatoes. These waiting periods take into account the effect of sun, wind, and rain on the chemicals while they are in the garden so that the harvested product is virtually free of the chemical at harvest. This is the reason you cannot harvest first and then wait the required number of "days to harvest" with the tomatoes safely on the kitchen cabinet. A week in the house is not comparable to a week in the garden.

Organic sprays fall into a number of categories—insecticidal soaps; naturally occurring substances like sulfur, copper, or potassium bicarbonate; beneficial or pathogen-competitive organisms; plant oils; and plant extracts.

Insecticidal soaps were one of the first organic sprays to be advocated. The soaps help wash off the pests, especially delicate insects like aphids, and in some cases they prove toxic to the pest—interfering with breathing or the protective covering on the insect. They also can be toxic to the tomato plant if used when the temperature is too hot or if the concentration is too strong. The key seems to be to test first by spraying a plant and observing it for a few days, spray in the early morning or evening when the temperature is cooler and the sun is less intense, and try a more dilute rate of application. The label typically recommends a range of dilutions. Start with the lowest rate first.

Naturally occurring materials such as sulfur, copper, and potassium bicarbonate have been used for decades. They may not work as spectacularly as chemical pesticides, but they usually can be used to good effect with some persistence. Sulfur, for example, is a good miticide and a fair fungicide. It has a tendency to burn the foliage in really hot weather (don't use at all on squash, cucumbers, and other cucurbits because it will burn the foliage), but it can usually be used in the cooler parts of the day. Alternating it with neem oil has given good mite control in the author's garden. Baking soda sprays using 1 tablespoon of baking soda (sodium bicarbonate) with 1 tablespoon of dormant oil and ½ teaspoon of liquid dish soap per gallon of water have been found to have fungicidal properties—especially for certain mildews. Commercial formulations use potassium bicarbonate, which is more effective on powdery mildew.

Biological pesticides really gained notoriety with the development of a bacterial spray using *Bacillus thuringiensis* (Bt). This organism is not toxic to people, pets, and most beneficial organisms. It works on caterpillars with an alkaline gut like the tomato hornworm and the tomato fruit worm. Unfortunately, it may also get a few butterfly larvae, but most don't frequent the tomato patch, so hopefully they will not be affected. *Bacillus subtilis* (Bs) strain QST 713 is marketed as Serenade and works as a fungicide by outcompeting the pathogens. Other versions include Kodiak, Sporatec, and Sonata. There are other strains of Bs that are used to control root pathogens on other crops.

Spinosad is a good example of a biological insecticide that was found by careful observation. An employee of the Lilly Company discovered it in an abandoned rum distillery on a Caribbean island in 1983. In the laboratory, it was found to be a metabolite of the actinomycete *Saccharopolyspora spinosa*. Spinosad has to be ingested by the target insect pest, so most natural enemies are not affected. Honeybees are rather sensitive, but damage can be minimized by applying the spray in the evening when the bees are not active. Spinosad breaks down within a few days, and the dried material has little effect on nontarget species.

Beneficial biological agents include mycorrhizal fungi that form a beneficial relationship with plant roots—they get food from the roots, and the roots get protection from pathogens and a vastly increased area of potential water/nutrient absorption from the fungal mycelia. Fungi, bacteria, protozoa, and beneficial nematodes all function together in a healthy soil to foster plant growth. In fact, the science of the soil web of living organisms can be rather daunting to the point that for most of us, keeping the compost handy is our main solution. A healthy supply of earthworms in the soil is said to be an indicator of good soil health, and obviously good plant growth is another.

Do you need expensive inoculants? This question remains unanswered. It would be foolish to waste money on inoculants if the soil is low in organic matter, poorly drained, or desert dry. If you amend the soil with compost, work it up into ridges for drainage, or build raised beds, then many of the beneficial organisms "will come." They get there in the compost and flourish in a well-managed soil. Similarly, a dry, sandy soil can be improved with organic matter and a good irrigation system. In low-rainfall regions of Texas, ridges won't be necessary. Perhaps the benefits of compost can be enhanced with the addition of selected microbes. For now, you will just have to experiment in your garden. Try to plant the same variety with and without the inoculants. Since these organisms can build up and move rapidly in the soil, it's a good idea to separate the test plants by 8 feet or more—at least, skip one row.

Botanical spray materials have proved a fertile area for development of organic sprays. Insecticidal oils like those of certain spices, such as cinnamon and clove, plus others like neem, sesame, and cedar oil have found their way into organic spray materials. Most have not had a great deal of university research to back up the claims made, but they are worth trying. These oils can be phytotoxic as formulated, or they may be even more likely to damage tomatoes when combined with other spray materials, so be sure to test them on a plant before applying them to the entire crop. Neem oil has the most extensive "track record" as an insecticide/fungicide/feeding suppressant, and its toxicity to people and pets is very low—in fact, it is even formulated into toothpaste. Other oils often labeled as "summer oils" are lightweight plant oils that also have had good results, but all should be used with caution and tested on a plant or two before being used on tomatoes.

Natural pyrethrums, rotenone, sabadilla, garlic sprays, and hot pepper have also been used in the organic spray tank. Pyrethrum and rotenone are often paired together—the pyrethrum has a fast knockdown, and the rotenone is a stomach poison. Sabadilla has some potential for controlling the true bugs like the stinkbug. It is a respiratory irritant once marketed as "sneezing powder," so it should be handled, like any pesticide, with care and attention to the label. Garlic is reported to run off the bugs and cure disease, so it is in a number of organic sprays. It does tend to be rather expensive, so you may want to get out the blender and make your own. Blend one bulb of garlic with 2 cups of water. Pour this mixture into a glass jar, cover with a lid, and let it sit for a day; then strain the liquid and add it to a gallon of water to make a spray mixture for aphids and similar insects. This will last about a week, perhaps less in the heat of summer. Hot pepper sprays are supposed to run off the rabbits—Texas rabbits may actually like them! It does spice up your tomatoes though.

Other pest control materials such as diatomaceous earth—the plant-pest type with the sharp points intact, not the pool-filter version— can be used to some benefit. It doesn't have any staying power, as it washes away quickly, but when dusted directly on an insect like the leaf-footed bug, it hastens the insect's demise. When the bugs are thick, you may want to treat every day or two.

POLLINATION

There has been a lot of concern about the health and welfare of honeybees—why are there no bees in the city? Is a disease killing them? Bees are not important for tomato pollination because tomato flowers are primarily wind pollinated. Cold, rainy days and, later in the summer, temperatures that are consistently above 75 degrees Fahrenheit at night and above 95 degrees during the day can cause a lot of tomato blooms to drop off. Blooms may not be getting good pollination, but that is not the main problem—it's just too hot for good fruit set. Small-fruited varieties and varieties with heat tolerance overcome this problem to some extent. Large-fruited varieties like Brandywine have large flowers that may need a helping hand to improve fruit set. A little cotton swab action may be just the ticket.

The Brandywine tomato and many other large-fruited tomatoes have a large flower that may be difficult to pollinate. On warm, sunny days using a cotton-tipped swab may help the process.

TOMATO-SNATCHING CRITTERS

Even the family dog is a suspect when tomatoes go missing. Of course, birds like to peck away at tomato fruits, and field mice or rats have been known to take their share. Uninvited two-legged critters sometimes help themselves—especially in a community garden. The old "Pesticides in Use" sign sometimes works, but your community garden likely won't allow it—even if it's not true. Wrapping fiber row cover around a cluster you want to photograph or brag about is effective for a while. Just use clothespins to secure the gauzelike material, which is easily removed for pictures or eventually for harvest.

WEEDS

Weeds in the small tomato patch or in container-growing situations are a minor problem. When you get to a larger planting—say, fifty plants or a market garden patch of an acre or more, then they can take over just as you begin to "count your chickens," so to speak. Those little grass sprouts can become a "walkway strangling barrier" to harvest if you don't act early. They also cut the harvest short as they begin to get more than their share of the water and nutrients.

One of the easiest ways to keep weeds down is to place a layer of newspapers eight to ten sheets thick (wet them first, and they won't blow away while you go for mulch) between the rows. You could even put them down solid, and cut holes for the tomato transplants. Then cover the paper with organic mulch like alfalfa hay or free wood chips from the local tree trimmer. Be sure to sprinkle nitrogen-containing fertilizer over the mulch several times during the season so that the paper is impregnated with nitrogen and the decomposition process is under way when it is time to work the mulch into the soil for the next garden season. If you use wood chips, you may need to add extra nitrogen to compensate for the nitrogen tie-up caused by microorganisms working on this high-carbon food source. If the wood chips are really deep at the end of the season, it would be best to rake them off and use them again next year.

When the tomato patch gets this weedy, you might want to consider starting over.

You could use a weed-barrier material under the mulch, but it would need to be removed before amending the soil for the next garden since it does not break down. Unfortunately, these woven synthetics often get fouled with Bermuda grass and other weeds by the end of the growing season, making them really tough to get up.

Don't forget the hoe. Hoes were discussed in more detail in the section on garden tools, but a scuffle hoe and a chopping hoe kept sharp can remove a lot of weeds quickly. Hand pulling weeds does not have a lot of advocates, although weeds always come up inside the tomato cage that defy the hoe. A good pair of gloves and some knee pads might prove invaluable. One of the reasons gardeners buy a tiller (the mini-tiller works great for cultivation) is to turn under weeds and cover crops. If you have a big garden, then mechanical cultivation is part of your weed-control arsenal.

Herbicides are tricky to use in the home garden. Products like glyphosate (Roundup, Eraser, systemic herbicides, and other generic formulations are now available) that kill most any plant they contact but don't hurt the soil are tempting to use when you are faced with a lot of tenacious weeds like Bermuda grass and nutsedge. You cannot use this material while the crop is growing, so it needs to be sprayed a season ahead of planting. Pre-emergence herbicides such as Treflan (trifluralin) that can be incorporated in the top 2 to 3 inches of soil at planting time can be counted on to prevent a number of weed seeds from germinating. A quick search of the Internet will show that there are many herbicides used for commercial tomato production, but few of them are recommended for use in the home garden. Tomatoes are extremely sensitive to many chemicals—even the organic ones like insecticidal soaps and oils—so it pays to be cautious and test any spray on or around a plant or two before treating the entire planting.

Acetic acid (vinegar) has been used with some success in home gardens. The kitchen version is not concentrated enough; you need a product with 15% to 20% acetic acid. This acid is strong enough that you need to be careful handling it—especially protect your eyes with safety glasses and wear elbow-length rubber gloves. A commercial preparation with clove oil added was rated highly by some gardeners who used it, but most noted it was expensive for the area covered and did not kill all the weeds—it works best on broadleaf plants, not so well on grasses or sedges. Unfortunately, these acetic acid–based herbicides are nonselective and would certainly burn tomato plants. So be sure to protect non-target species with a piece of cardboard when spraying.

7 Tomato Relatives

It would be a rare tomato lover who didn't enjoy some of the tomato kinfolk—peppers, eggplants, tomatillos, and potatoes are the most obvious. The tomato also has some "black sheep" relatives like bull nettle and the many solanaceous weeds that plague our pastures and cattle lots. Suffice it to say that there are toxins in even the edible species but not in the parts we consume. For example, tomato foliage isn't good for you, but for most folks the fruits are delicious and nutritious. Potato tubers are good, but don't eat the small green fruits that occasionally form. Generally the foliage of solanaceous plants is not edible, and avoid the entire plant of weed relatives.

PEPPERS

Peppers come in a great variety of colors, shapes, sizes, and flavors. Perhaps the easiest way to divide them is into hot peppers and sweet peppers. Most households don't need a lot of either category, but some, like the NuMex chiles or Ramshorn peppers, are great to put up in the freezer for use all winter long, and a few bags of chopped-up and frozen bells are nice to have for flavoring. Quite a few Texas households consider a few jars of pickled jalapeños a necessity. Some chiltepin pepper sauce or a shelf full of home canned taco sauce makes the winter more pleasant while we wait for the next crop of tomatoes and peppers to come in. Peppers cannot be planted into the garden as early as tomatoes—they can be stunted by cold weather—so wait for at least four weeks after the average last frost date before setting them out in the garden (March to April in most areas of Texas). Then they need heavy fertilization—almost to the point of burning them. The theory is to get as much early production as possible before it gets too hot. The plants will usually survive the heat of summer and even produce a few small peppers, and then they will come back for a good fall harvest.

Mildly hot or spicy peppers include most of the NuMex green chiles like Big Jim (10, 13, 14, 20) and others like Fajita Hybrid (20)—a bell with a bit of sizzle—along with Mariachi Hybrid (2, 6, 7, 16, 19, 20) and the slightly hotter Mexibel (20). Some of the mild jalapeños such as Fooled You (6, 8, 19, 20) and Señorita (17, 19, 20) could be included.

Seriously hot peppers include Cayenne (widely available), Chiltepin or Bird's Eye (8, 10, 11, 19, 20), Mucho Nacho (2, 7, 19, 20) or El Jefe (3, 16) jalapeños, Hungarian Yellow Wax (1, 3, 6, 8, 10, 13, 15, 16, 19, 20, 22), Inferno hot banana (21), Scotch Bonnet (19), Serrano (widely available), Tabasco (1, 6, 8, 10, 14, 19, 20, 21, 22), Yellow Aji (6, 10, 11, 13, 14, 19), and, of course, the habanero (widely available), which are guaranteed to "light your fire." Got to have the hottest? Look for the Indian (Assam) Naga Jolokia (search Amazon, eBay, and specialty pepper seed sources)—supposedly twice as hot as the hottest habanero. It is also called Bhut Jolokia or the Ghost pepper since you may run off into the sunset and disappear after eating one. This pepper is also known in Bangladesh, Sri Lanka, and Pakistan.

Sweet non-bell peppers such as Carmen (2, 3, 16, 19, 20), Red Marconi (1, 11, 19), and Spanish Spice (19, 20) are typical of the long sweet peppers that are great to use fresh, grilled, or fried. In fact, they are

Hot peppers come in a variety of shapes, sizes, and degrees of heat measured by Scoville units. Jalapeño, Padron, and habanero are represented here.

This large New Mexico chile pepper is just spicy enough to tease the palate and is rich with Southwestern chile flavor.

guaranteed to make a Philly cheesesteak sandwich come to life Texas style. If these sweet peppers don't get it done, then toss in a little attitude with some jalapeño slices. Grill some Texas Supersweet onions and thinly sliced venison, add mayo, and layer plenty of provolone cheese on a crusty sourdough roll, and you won't find a better "Philly" sandwich on the entire East Coast.

Sweet bells have the mildest flavor. When in doubt, plant the early hybrids. New varieties of peppers come out almost as frequently as tomatoes, so you have to be light on your feet and just know what to look for. The colored bells, especially the red ones, aren't that great for Texas gardens because they take so much time to mature. Try the yellow bells like Golden Summer (7, 19, 20, 22) or Lilac Hybrid (19, 20) (looks better on

The Mariachi pepper is mildly hot and super productive. It would even be
gorgeous as a background plant in the flower border.

The Senorita jalapeño is so mild you may think you're eating a bell pepper.
Chomp down a few at the next cookout, and you'll gain respect.

El Jefe is a medium-hot jalapeño that is very productive and has been bred for less skin cracking and scarring.

Serrano peppers make a great fresh salsa. For a little variety, chop them with onions, garlic, and cilantro spritzed with some lime juice and sprinkled with a little seasoned salt—great with fajitas, and your friends who don't like fresh tomatoes will appreciate the gesture. Put tomatoes in everything else though.

Need to light your fire?
Try the habanero.

The long, sweet to slightly pungent
Ramshorn peppers are easy to grow.
They are popular fried as a side dish
or as the basis for a wonderful Philly
cheesesteak sandwich.

Most colored bell peppers are not as productive because you have to wait longer to harvest, but Golden Summer grows very well in Texas gardens.

Gypsy does not have the typical bell shape but is extremely productive of pale yellow sweet peppers.

the bush than in a salad, but very edible). Most bells will eventually turn red if left on the bush long enough, but it will cost you in yield. Some tried-and-proven bells include Bell Boy Hybrid (16, 19, 20, 22), Big Bertha Hybrid (7, 8, 19, 20, 22), Camelot Hybrid (16, 19, 20), Colossal Hybrid (7, 20, 21), Gypsy Hybrid (6, 18, 19, 20), Peto Wonder Hybrid (19, 20), and Socrates Hybrid (16, 19).

EGGPLANTS

Even though most families won't need a lot of eggplants, there are plenty of fun varieties to try in the garden. If the spider mites don't get them in the summer, they will come back just like the peppers for a fall harvest. The white ones are pretty, although the skins seem a bit tough. Actually the green ones look ugly in the garden, but they look pretty when cooked and have tender skins and mild flesh without an overabundance of seeds. Louisiana Long Green (1, 10, 13, 15) is somewhat available in seed catalogs—Green Goddess (6, 19) and Raveena (3, 7, 8) are similar. There is also a standard-shaped green eggplant that is good. It seems to be popular in Louisiana, but I have not seen it listed in any seed catalogs (Jilo Tingua Verde Claro from Baker Creek Heirloom Seeds, source no. 1, appears to be similar). The Harris County Master Gardeners save seed each year from the standard Louisiana green eggplants (sold as "Oblong Green" in Louisiana nurseries) and grow transplants for their spring tomato/pepper plant sale.

Marbled white/purple ones are gorgeous, and they "eat good" too. Look for Antigua (19), Fairy Tale Hybrid (2, 6, 7, 16, 17, 19), the Italian heirloom Listada de Gandia (1, 10, 11, 15, 19), or Zebra Hybrid (3, 16, 19). The long, skinny, purple ones are mild with few seeds and tender skin. Hansel Hybrid AAS (16, 19—All America selection 2007), Machiaw Hybrid (19), and Ping Tung Long (1, 4, 8, 10, 15, 19) are excellent. Another excellent Italian heirloom from Tuscany, Prosperosa (12, 19), is roundish with white highlights around the calyx. Some, like the old Black Beauty (widely available), are notorious for being bitter when grown under stress in the summer. Always pick eggplants while the skin is glossy. Otherwise, they may be bitter with hard seeds.

Not sure how to use them in the kitchen? It's hard to beat slicing them ¼-inch thick, dipping them in an egg/milk mixture, and then covering them in Italian breadcrumbs. Fry them to a light brown on each side, and sprinkle with creole seasoning salt. Another good way to prepare them is in an Italian casserole including eggplant, Italian cooking sauce, breadcrumbs, and lots of gooey mozzarella, provolone, and flavorful parmesan.

TOMATILLOS

Tomatillos can be frustrating to grow since they often fail to set good crops. The most successful variety for the author has been Toma Verde (1, 3, 7, 10, 13, 15, 21, 22), a Ping-Pong ball–sized variety that produced an abundance of fruit in a small bed. Perhaps some of the other varieties are more dependent on cross-pollination. Other varieties to try include Purple Tomatillo (1, 3, 8, 11, 13, 17), with a sweet-tart flavor, and San Juanito (16), a determinate plant with large fruit. If you don't keep all the fruit harvested, this tomato relative can become

Louisiana Green eggplants may look strange in the garden, but they have tender skins, comparatively few seeds, and freedom from bitterness.

(left)
This small Antigua eggplant looks like it is made of marble. Not just a looker, it is productive and delicious.
(right)
The long purple eggplants like Ping Tung Long tend to be mild with few seeds.

The Prosperosa eggplant makes a delicious and productive addition to the tomato patch.

The Toma Verde tomatillo has been the most productive variety in the author's garden.

the new weed in your garden. When harvesting, feel the husk-covered fruit to make sure it is full size, since the husk develops fully long before the fruit does.

POTATOES

This is technically a warm-season crop, but we plant potatoes as soon as possible in the spring—two to four weeks before the last average frost (around Valentine's Day for many Texas gardeners). Like the tomato, this vegetable is from the high-altitude tropics and would like to have 70-degree days and 40-degree nights. The foliage can still be damaged by freezing temperatures, so be ready with the frost blanket to cover the foliage so it isn't blackened by a late frost. If you wait too late to harvest in the spring (typically after May–early June), the soil has warmed up too much and the spuds may be trying to sprout.

From south-central Texas and farther south, potatoes make a good fall crop. The trick is to save some of the small potatoes from your spring

Digging these Red La Soda potatoes is like having an Easter egg hunt. If dried without overcleaning and stored in single layers in a cool garage, they will last for months.

harvest and plant them whole in mid-August. They will mature under cool fall conditions, ensuring top quality and an excellent harvest to last through the winter. If you buy seed potatoes in late summer, they often fail to sprout because they have not gone through enough dormancy. You don't want to cut them into pieces because the warm soil in August will make them more likely to rot.

The varieties you usually find at a nursery or feed store are Kennebec for white potatoes and La Soda for red. There are a number of new varieties and unusual potatoes like purple ones or the fingerling types, but it is typically hard to get them early enough for Texas planting dates from mail-order sources. If you try to order some, specify shipping in time for the recommended planting date—even if you have to pay for overnight shipping.

Save your small potatoes from the spring harvest, and plant the whole potatoes in mid- to late summer for an even more productive fall crop.

The crescent or fingerling potatoes like this Austrian Crescent are delicious and don't turn to mush when cooked.

8 Tomato Source List

This list is not meant to imply any endorsement either by inclusion or exclusion. No doubt, there are many sources for tomato seeds and plants—not the least of which is a local nursery. Today it is common to find the uncommon in local nurseries. Folks like to plant heirlooms as well as the latest and greatest hybrids. For many tomatophiles, growing the biggest (not advised) and best (very much advised) is a serious competition. Personally, I would rather have 50 pounds of production from a single plant than one 2-pound Brandywine tomato, but if your garden is big enough, do it all!

The pictures in the following catalogs look oh so good, and the descriptions indicate every one of them is delicious—come to think of it, have you ever seen a tomato description in a catalog that read something like this: "Looks great but rather tasteless and grainy textured"? Wouldn't sell many seeds, would it? So presume that all of the catalog authors are optimists.

These addresses were correct at the time this book was written; the author is not responsible for changes of address or discontinued firms or varieties.

1. Baker Creek Heirloom Seeds, 2278 Baker Creek Rd., Mansfield, MO 65704
   www.rareseeds.com; Phone: 417-924-8917
   Free catalog. No hybrids or genetically engineered varieties here, but this beautiful catalog has homespun vegetables from all over the world—sort of a commercial "Seed Savers." *The Heirloom Gardener* publication is fascinating and informative, too.

2. Harris Seeds, 355 Paul Rd., PO Box 24966, Rochester, NY 14624-0966
   www.harrisseeds.com; Phone: 800–544–7938; Fax: 877-892-9197
   Free catalog, which includes growing tips for most vegetables; features "Customer Favorite" varieties and many commercial varieties. Professional grower catalog available.

3. Johnny's Selected Seeds, 955 Benton Ave., Winslow, ME 04901-2601
   www.johnnyseeds.com; Phone: 877-564-6697; Fax: 800-738-6314
   Free catalog. Many heirloom and open-pollinated tomato varieties as well as hybrids. Also a source for tomato rootstock varieties if you decide to graft your own. This is a fun catalog with good prices and service. Good variety of row covers available.

4. Kitazawa Seed Company, PO Box 13220, Oakland, CA 94661-3220
   www.kitazawaseed.com; Phone: 510-595-1188; Fax: 510-595-1860
   Specializes in Asian vegetable seeds. Japanese tomato varieties like Odoriko and Momotaro are quite productive in Texas gardens.

5. New Dimension Seed, Dimension Trade Company, PO Box 1294, Scappoose, OR 97056
   www.newdimensionseed.com; Phone: 503-577-9382;
   Fax: 503-543-4690
   Free price list. Specializes in Asian vegetable seeds—not too many tomatoes, but lots of other veggies to try.

6. Nichols Garden Nursery, Old Salem Rd. NE, Albany, OR 97321-4580
    www.nicholsgardennursery.com; Phone: 800-422-3985;
    Fax: 800-231-5306
    Great source for herbs and Asian and European vegetables.

7. Park Seed Company, 1 Parkton Ave., Greenwood, SC 29647-0001
    www.parkseed.com; Phone: 800-845-3369; Fax: 800-275-9941
    Free catalog. Good service and quality.

8. Pinetree Garden Seeds, PO Box 300, New Gloucester, ME 04260
    www.superseeds.com; Phone: 207-926-3400; Fax: 207-926-3886
    Free catalog filled with interesting varieties—some sectioned by
    region of origin (Asian, Italian, etc.). Catalog also includes an
    extensive selection of books. Small seed packets available at a
    reduced price.

9. Renee's Garden, 6116 Hwy. 9, Felton, CA 95018
    www.reneesgarden.com; Phone: 888-880-7228;
    Fax: 831-335-7227
    Source for gourmet vegetable varieties.

10. John Scheepers Kitchen Garden Seeds, 23 Tulip Drive, PO Box 638,
    Bantam, CT 06750-0638
    www.kitchengardenseeds.com; Phone: 860-567-6086;
    Fax: 860-567-5323
    Free catalog. Gourmet vegetable, herb, and flower varieties.

11. Seed Savers Exchange, 3094 North Winn Rd., Decorah, IA 52101
    www.seedsavers.org; Phone: 563-382-5990; Fax: 563-382-6511
    Specializes in heirloom varieties.

12. Seeds from Italy, PO Box 149, Winchester, MA 01890
    www.growitalian.com; Phone: 781-721-5904, Fax: 612-435-4020
    Great source for new and hard-to-find Italian and European
    varieties. The tomatoes have not been tested extensively in Texas,

but they need to be. Be sure to try the Verde da Taglio chard for a mild and delicious year-round green—you can't plant the entire garden in tomatoes!

13. Seeds of Change, c/o Marketing Concepts, PO Box 152, Spicer, MN 56288
www.seedsofchange.com; Gardening Hotline: 888-762-7333;
Fax: 320-796-6036
Source for organic flower, herb, and vegetable seeds. Also has a nice selection of gardening accessories and books.

14. South Carolina Crop Improvement Association, Foundation Seed Program, 1162 Cherry Rd., PO Box 349952, Clemson, SC 29634
www.clemson.edu/seed; Phone: 864-656-2520
Varieties developed for South Carolina (great for Texas, too), including many heirlooms. If southern peas grace your table almost as often as tomatoes, you need this catalog!

15. Southern Exposure Seed Exchange, PO Box 460, Mineral, VA 23117
www.southernexposure.com; Phone: 540-894-9480;
Fax: 540-894-9481
Catalog: $2.00. The catalog has a number of unusual and heirloom varieties with a Southern slant.

16. Stokes Seeds, PO Box 548, Buffalo, NY 14240-0548
www.stokeseeds.com; Phone: 800-396-9238; Fax: 888-834-3334
Free beautiful color catalog, and many of the varieties are great in Southern gardens.

17. Territorial Seed Company, PO Box 158, Cottage Grove, OR 97424-0061
www.territorialseed.com; Phone: 800-626-0866;
Fax: 888-657-3131
Specializes in gourmet varieties, heirlooms, and organic products.

18. Thompson & Morgan, 220 Faraday, Jackson, NJ 08527-5073
www.tmseeds.com; Phone: 800-274-7333; Fax: 888-466-4769
Free catalog. This is an English company with a U.S. distributor, so
the seed is rather expensive, but the company offers a number of
varieties you won't find elsewhere.

19. Tomato Growers Supply Company, PO Box 60015, Ft. Myers, FL 33906
www.tomatogrowers.com; Phone: 888-478-7333;
Fax: 888-768-3476
Lots of tomato varieties and an extensive list of sweet and hot
peppers. No self-respecting tomato lover would be without this
catalog or the next one—read them and drool!

20. Totally Tomatoes, 334 W. Stroud St., Randolph, WI 53956
www.totallytomato.com; Phone: 800-345-5977; Fax: 888-477-7333
Another specialty catalog with loads of tomatoes and peppers.

21. Twilley Seeds, 121 Gary Rd., Hodges, SC 29653
www.twilleyseed.com; Phone: 800-622-7333; Fax: 864-227-5108
Free catalog. Good selection and quality with a number of exclusive
varieties.

22. Willhite Seed Inc., PO Box 23, Poolville, TX 76487
www.willhiteseed.com; Phone: 800-828-1840; Fax: 817-599-5843
Free catalog. This Texas company started as a melon seed supplier
and now has a wide range of vegetable varieties adapted to Texas,
plus wildflowers and more.

Garden tools and supplies:
Gardener's Supply Company, 128 Intervale Rd., Burlington, VT 05401
www.gardeners.com; Phone: 800-427-3363; Fax: 800-551-6712

Seeds from Italy
2008 Catalog
Taste the Difference

SEEDS

Call: 800-622-7333 • Fax: 864-227-5108 • Visit: www.Seedexchange.com

SEED SAVERS EXCHANGE®
PASSING ON OUR VEGETABLE HERITAGE

celebrating
25 years

.com

Seed catalogs arrive in abundance during the winter while we're dreaming about next year's garden. Don't forget that local nurseries and feed stores have a nice assortment of seeds and transplants adapted to your area.

Index

Notes

Notes

Notes